AIRCRAFT SYSTEMS INTEGRATION OF AIR-LAUNCHED WEAPONS

Aerospace Series List

Design and Analysis of Composite Structures: With Applications to Aerospace Structures, Second Edition	Kassapoglou	April 2013
Aircraft Systems Integration of Air-Launched Weapons	Rigby	April 2013
Design and Development of Aircraft Systems, Second Edition	Moir and Seabridge	November 2012
Understanding Aerodynamics: Arguing from the Real Physics	McLean	November 2012
Aircraft Design: A Systems Engineering Approach	Sadraey	October 2012
Introduction to UAV Systems, Fourth Edition	Fahlstrom and Gleason	August 2012
Theory of Lift: Introductory Computational Aerodynamics with MATLAB and Octave	McBain	August 2012
Sense and Avoid in UAS: Research and Applications	Angelov	April 2012
Morphing Aerospace Vehicles and Structures	Valasek	April 2012
Gas Turbine Propulsion Systems	MacIsaac and Langton	July 2011
Basic Helicopter Aerodynamics, Third Edition	Seddon and Newman	July 2011
Advanced Control of Aircraft, Spacecraft and Rockets	Tewari	July 2011
Cooperative Path Planning of Unmanned Aerial Vehicles	Tsourdos et al.	November 2010
Principles of Flight for Pilots	Swatton	October 2010
Air Travel and Health: A Systems Perspective	Seabridge et al.	September 2010
Unmanned Aircraft Systems: UAVS Design, Development and Deployment	Austin	April 2010
Introduction to Antenna Placement and Installations	Macnamara	April 2010
Principles of Flight Simulation	Allerton	October 2009
Aircraft Fuel Systems	Langton et al.	May 2009
The Global Airline Industry	Belobaba	April 2009
Computational Modelling and Simulation of Aircraft and the Environment: Volume 1 Platform Kinematics and Synthetic Environment	Diston	April 2009
Handbook of Space Technology	Ley, Wittmann Hallmann	April 2009
Aircraft Performance Theory and Practice for Pilots	Swatton	August 2008
Aircraft Systems, Third Edition	Moir and Seabridge	March 2008
Introduction to Aircraft Aeroelasticity and Loads	Wright and Cooper	December 2007
Stability and Control of Aircraft Systems	Langton	September 2006
Military Avionics Systems	Moir and Seabridge	February 2006
Design and Development of Aircraft Systems	Moir and Seabridge	June 2004
Aircraft Loading and Structural Layout	Howe	May 2004
Aircraft Display Systems	Jukes	December 2003
Civil Avionics Systems	Moir and Seabridge	December 2002

AIRCRAFT SYSTEMS INTEGRATION OF AIR-LAUNCHED WEAPONS

Keith A. Rigby
BAE Systems, UK

A John Wiley and Sons, Ltd, Publication

This edition first published 2013
© 2013, John Wiley & Sons Ltd.

Registered Office
John Wiley & Sons Ltd, The Atrium, Southern Gate, Chichester, West Sussex, PO19 8SQ,
United Kingdom

For details of our global editorial offices, for customer services and for information about how
to apply for permission to reuse the copyright material in this book please see our website
at www.wiley.com.

Library of Congress Cataloging-in-Publication Data

Rigby, Keith A.
  Aircraft systems integration of air launched weapons / by Keith A. Rigby.
    pages   cm
  Includes bibliographical references and index.
  ISBN 978-0-470-97118-5 (cloth)
1. Air weapons.   2. Air-to-surface missiles.   3. Airplanes, Military–Armament.   4. Airplanes,
Military–Design and construction.   5. Systems integration.   6. Aeronautics–Systems engineering.
I. Title.
  UG1270.R54 2013
  623.4′51–dc23
                                      2012047732

A catalogue record for this book is available from the British Library.

Set 10/12pt Times by SPi Publisher Services, Pondicherry, India

Contents

Series Preface

The Aerospace Series has concerned itself largely with the design of aerospace vehicles and their systems, comprehensively covering aspects of structural and system design in theoretical and practical terms. There has been reference to military aircraft types in the books of the Series, sometimes as developments of commercial aircraft for surveillance or transport roles, and other times as specific combat types. However, there has been no detailed consideration of one aspect that is quite specifically applicable to military types – the carriage and release of air launched weapons.

In this book, the author takes a systems engineering view of the weapon and platform as an integrated whole for both manned and unmanned aircraft. The importance of considering the integration of the weapon with the airframe and with the aircraft systems is stressed as it is vital to the achievement of a safe and successful launch with a high probability of target destruction. This aspect of integration is important to the introduction of precision weapons with a high degree of accuracy to reduce the incidence of collateral damage, as well as making best use of costly weapons.

It is important for engineers and designers to visualise the totality of a system in order to gain an understanding of all that is involved in the establishment of requirements and the certification process to arrive at a coherent design of vehicle and infrastructure. Understanding the impact of external weapons installation on aircraft performance and handling, and the needs of the weapon for navigation and attitude information, is key to understanding how to aim, fuze and release a weapon for maximum effect. This understanding enables developments in new aircraft types and new weapon types to be understood and adapted so that the most effective weapon system can be selected and developed for a particular aircraft in order to respond to changing threats.

This book is a comprehensive treatise on the subject of air launched weapons and will be of great value to all design engineers, support engineers, air crew and armourers working on armed military aircraft. The message is reinforced by the introduction of a worked example of integration of a smart weapon with the airframe. It also provides a good background to people who have an interest, professional or casual, in the design of aircraft weapon systems. It is worth noting that the book carefully avoids any areas of security classification, thereby making the book accessible to a wide audience.

Peter Belobaba, Jonathan Cooper and Allan Seabridge

Preface

For any military conflict where Western air forces are involved, the public is accustomed to media coverage of weapons being targeted against the enemy. Whether this is an Air-to-Ground missile being aimed at a particular window in a building or a smart bomb destroying a strategically located bridge, the public understand that in modern warfare precision weapons can provide surgical attack capabilities whilst minimising collateral damage and harm to non-combatants.

However, the terminal accuracy of a guided weapon significantly depends on its priming prior to release. Simply, the launch aircraft and weapon together form a complex system where the performance of each component is interdependent on the performance of the other.

It is unusual for a weapon to be designed specifically for operation with a single aircraft type; it will generally be designed to provide a particular military capability when operated with a range of aircraft. The Weapon Design Organisation will generally define an idealised set of requirements to be placed on the launch aircraft such that if they are satisfied, the weapon can achieve its specified performance. However, although the Weapon Design Organisation has responsibility for the design, development and certification of the weapon, generally, it is the Aircraft Design Organisation that has responsibility for the design, development and certification of the total aircraft and for certification of the aircraft/weapon combination. As the aircraft may not be able to satisfy the set of ideal requirements placed on it by the weapon, the terminal performance of the weapon may be degraded.

Whenever a weapon has to be integrated with an aircraft, there will be a need for the Aircraft and Weapon Design Organisations to collaborate to satisfy the needs of the Government agency (the Contracting Agency) which contracts for the integrated capability. Whilst this may bring many organisational interaction challenges, the pure engineering activities which need to be undertaken are many and complex.

For the purposes of this book, weapons integration is divided into systems integration activities and aeromechanical activities (e.g. covering aerodynamics, structures and the airborne environment). Whilst all activities must be undertaken to realise the certified integrated product, this book concentrates on the aircraft systems integration aspects of air-launch weapons integration. However, this still covers a complex series of activities which are multi-disciplinary in nature and which it is unlikely that a single engineer in an organisation would have a deep understanding of all.

This book aims to give an insight into the various aspects of aircraft systems integration including consideration of the various subsystems required for the successful control of a weapon, systems engineering principles as applied to an integration programme, safety and certification, and provides a worked example of the integration of a smart weapon with an aircraft. By covering the broad scope of aircraft systems integration of air-launched weapons, it is intended that engineers at every level in their career will find something useful, be it a revision of previous knowledge, gaining an insight into the future direction of weapons integration or understanding the extent of weapons integration activities for those new to the discipline.

Acknowledgments

As with any textbook, there will always be people who have helped the author by sourcing material, checking draft manuscripts and supplying images to illustrate the content. This book is no exception. I would therefore like to thank all those people in the United Kingdom and United States who have helped in some way, but especially the following.

Paul Ellis, Pierre Miles and Rod Robinson who provided invaluable comments on the various draft manuscripts; and Chris Ryding, Nick Guard, Geoffrey Lee and Thiery Wurtz for use of their photographs. There are also those individuals within BAE Systems Military Air & Information and elsewhere who supported the production and vetting of the final manuscript and secured permission to use other copyright images.

A large amount of material in the book has drawn on the work of the many experienced weapons integration practitioners that have contributed to standardisation efforts within the Society of Automotive Engineers Aircraft Systems & Systems Integration Committee and the various NIAG studies. There are far too many to mention by name, but you know who you are. Thank you all.

List of Abbreviations

°C	Degrees Celsius
1PPS	One Pulse Per Second
3-DOF	Three Directions of Freedom
6-DOF	Six Directions of Freedom
A	Ampere
AC	Alternating Current
ACMI	Air Combat Manoeuvring Instrumentation
ADO	Aircraft Design Organisation
AEIS	Aircraft/Store Electrical Interconnection System
AGE	Aerospace Ground Equipment
AIM	Air Intercept Missile
AIR	Aerospace Information Report
AL	Application Layer
ALWI	Aircraft, Launcher and Weapons Interoperability
ALWI-CI	Aircraft, Launcher and Weapons Interoperability – Common Interface
ANSI INCITS	American National Standards Institute International Committee for Information Technology Standards
AOP	Allied Operating Procedure
API	Application Programme Interface
APOS	Application to Operating System Interface
AS	Aerospace Standard
ASAAC	Allied Standard Avionic Architecture Council
ASI	Aircraft Station Interface
ASRAAM	Advanced Short-Range Air-to-Air Missile
AWFL	Airworthiness Flight Limitations
BC	Bus Controller
BIT	Built-In Test
BPSK	Bipolar-Phase Shift Key

C of D	Certificate of Design
C/A	Course Acquisition
C/N_0	Carrier to Noise Ratio
CCIP	Continually Calculated Impact Point
CEP	Circular Error Probability
CLARA	Common Launch Acceptability Region Approach
CMM	Critical Monitor Message
COTS	Commercial Off-the-Shelf
CSI	Carriage Store Interface
CSSI	Carriage Store Station Interface
D	Direct Interface (1D, 2D, 3D and 4D)
dB	Decibel
dBHz	Decibel-hertz
dBic	Decibels relative to an ideal, circularly-polarised isotropic antenna
dBK^{-1}	Decibels per Kelvin
dBW	Decibels (relative to 1 W)
dc	Direct Current
DC	Direct Current
DDP	Declaration of Design and Performance
Def Stan	Defence Standard
DMC	Display Mission Computer
DO	Design Organisation
DoD	Department of Defense
DoDAF	Department of Defense Architecture Framework
DRL	Data Requirements List
EBR-1553	Extended Bit Rate 1553
EMC	Electromagnetic Compatibility
F	Noise Figure
FAA	Federal Aviation Administration
FFA	Functional Failure Analysis
FMECA	Failure Modes Effect Criticality Analysis
FOM	Figure of Merit
FTA	Fault Tree Analysis
g	Gain
GASIF	Generic Aircraft–Store Interface Framework
G_e	Effective Gain
GHz	Gigahertz
GOA	Generic Open Architecture
GPS	Global Positioning System
GR	Ground Attack/Reconnaissance
GRAM	GPS Receiver Application Module
GSM	Generic System Manager
GWR	Guided Weapon Release
HB1	High Bandwidth 1
HDBK	Handbook
HE	High Explosive

HOTAS	Hands on Throttle and Stick
HOW	Hand-Over Word
HRI	Hazard Risk Index
HUD	Head-Up Display
Hz	Hertz
IAD	Interface Agreement Document
ICAO	International Civil Aviation Organisation
ICD	Interface Control Document
ICN	Interface Change Note
ICWG	Interface Control Working Group
IER	Interface Exchange Requirements
IIR	Imaging Infra-Red
IMB	Interface Management Board
IMM	Interface for Micro Munitions
IMU	Inertial Measurement Unit
INCOSE	International Council on Systems Engineering
IR	Infra-Red
IRIS-T	Infra-Red Imaging System Tail/Thrust Vector-Controlled
IRS	Integration Requirements Specification
IRS-CM	Integration Requirements Specification – Compliancy Matrix
IRST	Infra-Red Search and Track
IRS-VP	Integration Requirements Specification – Validation Plan
ITEA	Integrated Test, Evaluation and Acceptance
J/S	Jammer to Signal Ratio
JICWG	Joint Interface Control Working Group
JSP	Joint Service Publication
K	Kelvin
kg	Kilogram
kHz	Kilohertz
kts	Nautical Miles per Hour
kΩ	Kilohm
LAN	Local Area Network
LAR	Launch Acceptability Region
LGB	Laser Guided Bomb
LOAL	Lock-On After Launch
LOBL	Lock-On Before Launch
LSB	Least Significant Bit
LSP	Least Significant Part
m	Metre
mA	Milliampere
MAA	Modular Avionics Architecture
MAR	Military Aircraft Release
MASS	Master Armament Safety Switch
Mbits/s	Megabits per Second
MDA®	Model Driven Architecture
MDT	Mass Data Transfer

MHz	Megahertz
MiDEF	Mission Data Exchange Format
MIL-STD	Military Standard
MMHI	Micro Munition Host Interface
MMSI	Miniature Mission Store Interface
MOS	Module to Operating System Interface
ms	Millisecond
MSB	Most Significant Bit
MSCI	Miniature Store Carriage Interface
MSCS	Miniature Store Carriage System
MSI	Mission Store Interface
MSL	Module Support Layer
MSP	Most Significant Part
Mux	Multiplex Data Bus
NATO	North Atlantic Treaty Organisation
NIAG	NATO Industrial Advisory Group
NIU	Network Interface Unit
NUAI	NATO Universal Armament Interface
OFP	Operational Flight Programme
OS	Operating System
OSI	Open Systems Interconnection
OSL	Operating System Layer
P(Y)	Precision Code
PIM	Platform Independent Model
P_k	Probability of Kill
PPS	Precise Positioning Service
PRN	Pseudo Random Noise
PSI	Platform Specific Implementation
PSICD	Platform–Store Interface Control Document
PSM	Platform Specific Model
PTTI	Precise Time and Time Interval
PVT	Position, Velocity and Time
QoS	Quality of Service
REAC	Reaction Time
RF	Radio Frequency
RT	Remote Terminal
RTS	Release To Service
S&RE	Suspension & Release Equipment
SAE	Society of Automotive Engineers
SCM	Store Control Message
SDB II	Small Diameter Bomb II
SDM	Store Description Message
SIL	Systems Integration Laboratory
SMM	Store Monitor Message
SMP	Stores Management Processor

SMS	Stores Management System
SNR	Signal-to-Noise Ratio
SoD	Statement of Design
SPEAR	Selected Precision Effects At Range
SPS	Standard Positioning Service
SRR	System Requirements Review
STANAG	Standardisation Agreement
SV	Space Vehicle
T	Output Noise Temperature
t	Temperature
T_0	Ambient temperature
$T_{antenna}$	Antenna noise temperature
TC	Transfer Control
TC	Transfer Connection
TD	Transfer Data
TEACASE	Thermal Effects on Airborne Conventional Armament Stores and Equipment
TLE	Target Location Error
TLM	Telemetry Data Word
TM	Transfer Monitor
TMP	Time Mark Pulse
t_n	Noise temperature
TOO	Target of Opportunity
TRM	Technical Reference Model
TTFF	Time to First Fix
™	Trade Mark
UAI	Universal Armament Interface
UAS	Unmanned Air System
UAV	Unmanned Air Vehicle
UK	United Kingdom
UML	Unified Modelling Language
UPC	Unique Planning Component
US	United States (of America)
UTC	Universal Time Co-ordinated
V	Volt
VC	Virtual Channel
W	Watt
WBS	Work Breakdown Structure
WCP	Weapon Control Panel
WDO	Weapon Design Organisation
WGS-84	World Geodetic System – 1984
WOWS	Weight-Off-Wheels Switches
WSSU	Weapon Station Switching Unit
XOS	Extended Operating System
xUML	Executable Unified Modelling Language
μs	Microsecond

1

Introduction to Weapons Integration

1.1 Introduction

One of the key differences between civilian and military aircraft is that many military aircraft have the ability to carry and release weapons. From the earliest days of aviation where the pilot would drop simple bombs by hand, engineers have striven to develop the capability to accurately deliver weapons against targets reliably and safely. Today, for a successful target engagement, it is essential for the aircraft and weapon to be integrated such that the full capabilities of the weapon can be exploited. The release of a weapon whether it is a forward-fired missile or a downward ejected store such as a fuel tank, from either an externally mounted pylon or from an internal bay, creates issues such as the ability to achieve safe separation and the ability of the aircraft structure to withstand the imparted loads. The complexity of weapons integration is increased when the requirements for priming and aiming are considered. The integration of weapons onto aircraft therefore requires a multi-disciplinary set of capabilities within the integration organisation.

The generic term for any mission payload carried on an external pylon or an internal bay on a non-permanent basis is a 'store'. The family of stores includes weapons, fuel tanks, countermeasure pods and so on. Whilst many of the stores would only be released from the aircraft under emergency conditions (e.g. in the event of an engine failure during take-off), weapons are designed to be released as a matter of course.

This book gives an introduction to the subject of weapons integration, primarily from the viewpoint of aircraft electrical and computing systems integration, and explores the systems integration problem space, outlining the importance of systems integration and the contribution of industry standards in achieving an integrated aircraft and weapons capability.

The following sections outline the contents of the main chapters of this book and then introduce the various aspects of weapons, the subsystems they employ and the contributions

Aircraft Systems Integration of Air-Launched Weapons, First Edition. Keith A. Rigby.

these make to a successful target engagement. By gaining an understanding of weapons, the subject of their integration with the launch platform can be explored.

1.2   Chapter Summaries

1.2.1   The Systems Integration Process

As with any system design, a structured, top-down approach is essential. However, for weapons integration, the higher level requirements will also include aeromechanical aspects such as the desired release envelope, carriage life, the number of weapons that can be carried, influences of other weapon and store types to be carried on the same sortie, and so on. In a series of chapters, the book will outline the basic systems engineering process employed in a typical weapons integration programme. This will include the definition of an appropriate set of requirements and their partitioning to subsystems, through systems implementation, to qualification and certification. The types of requirements that should be considered will be discussed and the benefits of minimising the number of initial requirements and the need to avoid 'over engineering' of requirements will be explained.

The need to consider safety from the outset and its place in the overall systems integration process will also be covered as will the individual responsibilities of the aircraft and weapon design organisations (WDOs).

Once the top-level integration requirements have been defined, there is a need to decompose these into more detailed requirements and to then partition these to the aircraft subsystems. This segmentation process may use software-based requirements management tools as these will, in due course, assist in the validation and verification of the system implementation against the requirements. This partitioning exercise depends on the actual aircraft system architecture and will therefore differ between aircraft, with the requirements being partitioned to individual aircraft subsystems, which could be implemented in both hardware and software. Each requirement placed on the aircraft's subsystems will need to be proven for correct implementation, and the subsystems will then be progressively built up into an overall integrated system that provides the required military capability. This will include the testing of the system and its components in a Systems Integration Laboratory (SIL). Employing either weapon simulators or inert weapons with operational electronics, integration testing will test that all the aircraft subsystems are working together to control the weapon. Any problems which are identified would then need to be corrected during an iteration of the system design and implementation.

The typical individual responsibilities of the aircraft and WDOs will also be discussed.

1.2.2   Stores Management System Design

The first electrical systems to control the release of stores were based on relays that when energised would switch the current to the bomb rack, causing it to open. The relays were operated by the Bomb Aimer pressing the release button, thereby routing a current to the relay coils. From a safety and certification viewpoint, it is essential that an aircraft only releases a store when intended. This appears to be an obvious requirement, but it is the primary driver in the design of the armament system of an aircraft.

This requirement forced the design of systems with multiple breaks in the bomb rack firing chain such that a single failure, on its own, could not cause weapons to be released inadvertently. This basic principle is often referred to as the 'no single failure' criteria as no single failure can cause an unintended release of a store when not intended. A second 'availability' principle is often also quoted such that no single failure shall prevent a release when intended.

In a modern military aircraft, the simplicity of the first electrical systems has been replaced by a subsystem in its own right which is generally referred to as the Stores Management System (SMS). The SMS manages the weapon load-out and controls the safe arming, release, jettison and operation of any store loaded on the aircraft, including the generation of the high-integrity data messages required by modern smart weapons to ensure their safe operation. This increased level of functionality and the need to ensure that weapons are only released or jettisoned when required are the primary drivers in SMS design, adding complexity to the hardware implementation and introducing the need for embedded safety critical software. Chapter 4 will explore the SMS design considerations and outline common system architectures that are found in modern military aircraft. Design considerations for aircrew training for Air-to-Ground weapons delivery will also be discussed.

1.2.3 The Global Positioning System

The accuracy to which weapons can be delivered on target is a key requirement of most new weapon programmes. Since the first Gulf War in 1991, there has been an increased use of navigation technology such as the United States' Global Positioning System (GPS) to assist in the terminal guidance of weapons. This increased use of GPS receivers in guided weapons has enabled the continued potency of legacy aircraft. The integration of such so-called smart weapons brings with it special problems for the aircraft systems integrator. Chapter 4 will also outline the basic operation of GPS and discuss a number of aircraft system design issues that need to be considered when integrating such weapons.

1.2.4 Weapon Initialisation and Targeting

Different weapons demand different methods of targeting. Targeting a weapon, be it an Air-to-Ground weapon or an Air-to-Air missile, can be very complex and place great demands on the performance of the aircraft systems. For the accurate targeting of a smart weapon, it is essential that the aircraft and weapon axis reference systems are initialised to provide a common reference, thereby removing position and velocity errors. The chapter on weapon initialisation and targeting will discuss weapon initialisation and examine the different ways in which weapons are targeted covering the accurate delivery of ballistic bombs, the flexibility of targeting for smart weapons and the sensor types and target prosecution strategies of Air-to-Air missiles. Training for the delivery of Air-to-Air missiles will also be discussed.

1.2.5 The Role of Standardisation in Weapons Integration

From the earliest days of guided weapons, weapon designers have defined their interfaces to optimise the requirements against technologies that have been available. This has led to a plethora of different interfacing systems existing on aircraft.

In a modern aircraft/store integration programme, technical standards play a significant part in improving interoperability and reducing integration costs and timescales. For example, standards such as Military Standard 1760 (MIL-STD-1760) have sought with some success to reduce the number and variation of interfaces and thereby improve interoperability of different weapons across many platforms. With the success of interfacing standards, the weapons integration community has also developed standards that ease other areas of integration such as protocol standards that facilitate greater interoperability in aircraft computing systems. The chapters covering standardisation will review a number of important weapon interfacing standards including MIL-STD-1760 (Aircraft-Store Electrical Interconnection System), the Miniature Mission Store Interface Aerospace Standard (AS5725) and the standard for the Interface for Micro Munitions (AS5726).

Other important industry technical standards such as the Generic Aircraft-Store Interface Framework, the Mission Data Exchange Format and the Common Launch Acceptability Region Approach will also be covered.

1.2.6   Interface Management

Weapons integration programmes require the aircraft design organisation and the WDO to collaborate. The interface between the aircraft and the weapon must therefore be agreed by the two parties and documented in an Interface Control Document (ICD). The ICD defines all information relevant to the integration such as mechanical attachments, electrical signal sets, data structures, timeline (a detailed temporal sequence of data and state transitions required for the aircraft to operate the weapon), environmental data, aerodynamic data and so on. Negotiation of the ICD can be a significant activity, particularly when a new weapon is being developed in parallel with its integration with the platform.

The chapter on interface management (Chapter 8) will give an overview of the type of information that has to be agreed in an ICD but will also detail the differing approaches between the United States and Europe in controlling the interface data and managing the process to agree the ICD. Where a weapon is to be integrated across a number of platforms, there may be a need for several programmes to develop and agree ICDs simultaneously. This provides a significant organisational challenge. Strategies such as the need for an Interface Control Plan and multi-programme Interface Control Working Groups will also be considered. Chapter 8 will also discuss an effective management process for controlling and reducing integration risk.

1.2.7   A Weapons Integration Scenario

Having covered the various aspects of aircraft systems integration of air-launched weapons, two chapters will draw together all the proceeding chapters and consider how they would contribute to a 'real' weapons integration programme (albeit, based on a hypothetical weapon). The first chapter (Chapter 9) identifies example specifications and features for some aspects of the weapon and the aircraft which will be used to illustrate elements of the system's integration, by considering the activities relating to the design implementation. In order to aid the understanding of the scenario, a typical weapon loading (to the aircraft) to dispersion (safe separation from the aircraft) sequence will also be discussed.

Chapter 10 will continue to draw together the proceeding chapters of the book and expand on other areas of the integration programme, in particular, those activities relating to system proving and certification.

1.2.8   'Plug and Play' Weapons Integration

In an ideal world, it would be possible to bring together a weapon and aircraft that have no knowledge of each other. Both would employ standard mechanical attachments and a standard electrical interface. The aircraft/weapon combination would also have a standard method for both sides of the interface to discover each other and re-configure their operation such that the aircraft is able to exploit the capabilities of the weapon with no system or software changes. Re-configuration would be achieved by the exchange of data only. If this could be achieved, we would have what is commonly referred to as 'Plug and Play' weapons integration. However, the reality is far more complex, and the 'Plug and Play' question has been debated for several years.

In 1998, the North Atlantic Treaty Organisation (NATO) Air Group 2 initiated a study entitled Aircraft, Launcher & Weapons Interoperability (ALWI). The study had the remit to gather data on existing aircraft and weapons, identify where interoperability existed and outline what would be needed in the future to have increased interoperability. The ultimate aim of NATO was to achieve a 'Plug and Play' capability for the systems integration of weapons, although the study postulated that for such a capability to be realised, significant technology developments would be required. Two subsequent NATO studies relating to the advancement of the 'Plug and Play' concepts have paved the way for 'Plug and Play' weapons integration capabilities to be developed. However, the realisation of this capability is still the ultimate goal for aircraft/weapons integration. In a series of chapters, various technologies and concepts will be discussed relating to this goal.

In looking at the implementation of 'Plug and Play' weapons integration, it is also envisaged that open system architectures will provide the basis on which this capability will be fully realised. Chapter 12 on open systems will discuss how open architecture mission systems coupled with modular weapon function software will have a significant bearing on the future of weapons integration. The chapter will also consider the evolution of the contracting and industry environment and how this will pave the way for delivering open systems technology that supports the introduction of a 'Plug and Play' weapons integration capability, thereby reducing integration costs and timescales.

The first incarnation of 'Plug and Play' is the United States' Universal Armament Interface (UAI), a concept which aims to break the dependence on the aircraft software upgrade cycles thereby enabling new weapons to be introduced to service quickly. The chapter on UAI (Chapter 13) will outline the development of 'Plug and Play' concepts from the NATO studies through to the UAI concept and beyond and identify the benefits that this capability brings to weapons integration programmes.

1.2.9   Weaponised Unmanned Air Systems

The integration of weapons with manned aircraft is a complex but well-understood discipline. However, the advent of weaponised unmanned air systems brings new challenges for weapons integration. Several programmes have fielded so-called 'killer drones'. Chapter 14 on the weaponisation of unmanned systems will consider the differences between weaponised manned and unmanned systems and outline strategies for partitioning the overall system between airborne and ground-based segments.

1.2.10 Reducing the Cost of Weapons Integration

Having discussed the various elements that need to be brought together to have a successful systems integration programme, the final chapter will investigate the need to reduce the costs and timescales associated with aircraft/weapons integration programmes. Both industry and national procurement organisations agree that the costs associated with the integration of a new weapon with an aircraft is expensive and the time taken to field new capability is too long. The cost drivers are many, and for costs and timescales to be reduced, there is a need to address a multitude of factors. This chapter will identify the cost drivers and the initiatives required to improve business efficiencies such that weapons integration programmes can be streamlined, thereby reducing integration costs. The final chapter will also argue that even in a climate of reducing defence budgets, efficient weapons integration programmes can be a source of increased business for defence companies.

1.3 Weapons

1.3.1 Types of Weapon

A key differentiator between a weapon and other types of stores is that weapons are designed to be released as a matter of course in order to prosecute a successful mission. To enable the aircraft systems integration problem space to be explored, it is first necessary to have a basic understanding of weapons and their subsystems. In the remainder of this chapter, a range of weapon types will be introduced including simple ballistic bombs, Air-to-Air missiles and modern 'smart' weapons and will identify the primary requirements of a weapon. In doing so, we will identify how these requirements generate the need for individual weapon subsystems and the demands these place on the launch aircraft. Multi-weapon carriage systems will also be considered as the means to increase weapon load-out, albeit by increasing the complexity of the systems integration activities.

Having a basic understanding of the weapons themselves and their requirements will enable the aircraft systems integration problem space and the differences in the integration of each weapon type to be explored in more detail in subsequent chapters. However, before considering weapon requirements, it is first necessary to understand the typical types of target that the weapon will have to defeat. Understanding this will enable the primary requirements for a weapon to be identified. It is those basic requirements which drive the design of the weapon including the need for the primary weapon subsystems.

1.3.2 Targets

The primary aim of an air-launched weapon is to cause an effect on its target (for the purposes of this chapter, it is assumed that the effect is to destroy the target by the energy released during the weapon's impact). In a military conflict, there are many types of potential targets such as troops, buildings, vehicles, other aircraft, ships and airfields.

The successful destruction of these targets requires the employment of different warheads and delivery methods. For example, in the past, weapons such as cluster bombs were employed against troops. Whilst effective against people, a cluster bomb is all but useless against a hardened building, where a bomb with a level of penetration capability and a blast warhead would be more effective. Engagement of airborne targets also requires a delivery mechanism such as a homing missile for engagement.

An examination of the different types of weapon which can be found in the inventories of the world's air forces will show that ballistic bombs, bombs with added control mechanisms that improve accuracy and powered weapons having a greater stand-off range (including Air-to-Air and Air-to-Ground missiles) exist, each to be used against different targets.

However, before delving into specific weapons more deeply, it is important to remember that weapons only exist to satisfy a military need. It is this which ultimately dictates the design and configuration of any weapon, and hence its integration complexity.

1.3.3 Weapon Requirements

For any weapon, there are three basic requirements. These are its lethality against its intended target, the accuracy (precision) with which it can engage the target and the stand-off range from which it can be launched against the target. From these three basic requirements, all the detailed design requirements of a weapon can be derived, detailed requirements which in turn drive the choice of weapon subsystems required to ensure a successful target engagement.

The following sections will discuss each of these requirements in turn and outline the weapon subsystems which are then derived from these requirements.

1.3.4 Lethality

1.3.4.1 Warheads

The achievable lethality of a weapon against a target will depend on the target type and the warhead used. Table 1.1 identifies the common types of warhead that could be used to defeat typical targets.

Of course, it would be possible to deploy a warhead against a target for which it was not designed to defeat. Whilst there may still be some effect on the target, this is unlikely to be optimised and raises the risk of collateral damage.

For greatest effect on the target, the warhead must be detonated at the right instant. For example, a shaped charge is most effective when it is detonated at a distance above the armour plate at which it is aimed. This is because the molten metal jet takes a finite time to build. If the warhead is detonated too early, the jet will have started to collapse before it comes into contact with the armour. If the warhead is detonated too late, then the jet will not have sufficient time to build. Both these cases could significantly reduce the efficiency of the weapon. Therefore, another key weapon component is the fuze or fuzing system.

1.3.4.2 Fuzes

Many different types of fuzing systems exist with probably the simplest being an impact fuze. Here, the rapid deceleration of the weapon is detected and a small pyrotechnic charge is initiated which itself initiates the main explosive charge of the warhead. For penetrating weapons,

Table 1.1 Warheads versus target types

Warhead type	Blast/frag	Shaped charge	HE penetrator	Remarks
Target type				
Troops	√			Note 1
Block wall buildings	√		√	Note 2
Hardened buildings/shelters/ buried targets			√	Note 2
Soft-skinned vehicles	√			Note 1
Missile launchers	√			Note 1
Armoured vehicles		√		Note 3
Other aircraft	√			Note 1
Radar installations	√			Note 1
Ships			√	Note 2
Airfields			√	Note 2

Note 1: Blast/fragmentation warheads combine a pressure wave with highly energetic shaped fragments.

Note 2: High explosive (HE) penetrator used to breach hardened structure before detonating and producing a powerful blast wave.

Note 3: A shaped charge generates a high-speed molten metal jet which penetrates armoured structure.

the fuze could be enhanced to incorporate a circuit that delays initiation. This enables the weapon to penetrate the target before detonating, thereby maximising damage. A third type of fuze is the crush fuze which is usually mounted on a frangible circuit board located at the front of some weapons. When impact is sensed by the initial impact breaking an electrical loop embedded in the card, the pyrotechnic chain is initiated.

As noted earlier, the effect of some types of warhead is maximised if they are initiated before impact. For this, some form of proximity sensor would be employed which uses radio frequency pulses or a pulsed laser to continuously measure the distance between the sensor and its target. For weapons requiring an air-burst, it is not uncommon for a simple radar altimeter to be employed as the sensing input to the fuze initiation circuits.

As different targets require different fuze operation and performance, modern weapon fuzes often have their functions programmable via the interface with the aircraft. Typical parameters that could be programmed are identified in Table 1.2.

1.3.5 Precision

When specifying a weapon, the accuracy with which it can engage its target is important. Ballistic, unguided bombs are notoriously inaccurate with miss distances of the order of tens of metres (the aiming of ballistic bombs will be covered in more detail in Chapter 5). By comparison, modern guided weapons can achieve precision engagements in the single metre or less range. A precision-guided weapon therefore enables warhead size to be reduced such that more of the explosive yield of the warhead is used to defeat the target, rather than it being wasted or being the cause of collateral damage.

Table 1.2 Typical programmable fuze parameters

Parameter	Remark
Fuze function delay from release	Time after the weapon has confirmed that it has been released from the aircraft before the fuze is allowed to detonate the weapon. This is required, for example, by weapons that have to function at a specific time after release (e.g. for a weapon that dispenses submunitions)
Fuze function delay from impact	Time delay after the weapon has detected that it has impacted the target. This is required, for example, to enable a weapon to penetrate its target before exploding
Void/layer number	Modern smart fuzes may contain accelerometers which can sense the retarding of the weapon. Many hardened military targets (e.g. command bunkers) are constructed of a number of layers interspersed with voids. The fuze may be programmed to detect the number of layers/voids through which it has passed before it detonates the warhead
Function time from event	Sets a time that the fuze must function after a specific programmed event has been detected. This could be, for example, an impact or when the fuze has detected that it has passed through a number of layers/voids

The need for precision therefore drives the need for a control system to manoeuvre the weapon. This is often coupled with a sensor that can detect the target and an autopilot to determine the projected miss distance and make control surface corrections to steer the weapon on to its target.

1.3.5.1 Sensors

Guided weapons employ many different types of sensors including those that can detect heat or reflected laser energy, imaging sensors (both video and Imaging infra-red (IIR)), radar and in-built navigation such as inertial systems (now usually supplemented by a GPS receiver).

Although some early weapons used video sensors, by far, the most successful of early guided weapons were those that had a sensor that could detect laser energy reflected from the target. The laser energy is generated, for example, by a portable system such as would be used by a forward air controller, or by a designator pod carried either by the launch aircraft or by a co-operating aircraft. The use of a laser designator pod (which itself, often has the laser function integrated as a part of a more complex electro-optical targeting pod) provides the aircraft with an excellent capability to detect and attack static targets. However, one major shortfall is that the laser has to be illuminating the target constantly during the engagement, and this is not always possible due to, for example, smoke or weather. For this reason, weapons that employ an inertial system coupled with a GPS receiver have been developed. This enables targets with a known location to be engaged when purely visual conditions are unfavourable. When conditions permit, the use of a laser to supplement the inertial/GPS improves end-game precision.

Air-to-Air missiles generally employ one of two types of sensors, these being infra-red (IR) and radar (although some missiles employ both sensor types).

Early Air-to-Air missiles employed a single-element IR sensor to detect the heat emitted from an enemy fighter's engines. This relatively simple technology would detect any source of heat including the sun or heat reflected from the launch aircraft's fuselage, thereby causing the missile to lock on to a false target. Although early seekers were improved to counter false targets, modern IR sensors now use IIR focal plane arrays. This enables the missile control algorithms to discriminate between a real aircraft target and false or spoof targets (e.g. IR countermeasure flares).

For longer range engagements, an IR sensor does not provide adequate performance. Radar sensors have a significantly greater range and therefore afford the ability for a fighter aircraft to increase its stand-off distance from its target. However, with the increased use of jamming as a defensive mechanism, radar sensors employ counter-countermeasures such as providing a 'home on jam' capability where the sensor acts in a passive mode to identify the relative bearing of the jamming signal, enabling the missile's guidance electronics to manoeuvre and home in on the source of the jamming signal.

A variation of 'home on jam' is to use a radio frequency receiver to detect the signals from an air defence radar, enabling the weapon to home in on to the transmitter.

Several Air-to-Ground weapons now employ active radar seekers operating in the millimetre wavelength range to detect their metallic targets.

1.3.5.2   Control Systems

Once a weapon knows where its target is, it needs to be able to steer itself so that it achieves a successful engagement. Guided weapons therefore need to have aerodynamic surfaces driven by actuators, which themselves are controlled by a system which is capable of translating information from the sensor into guidance commands. Some weapons have a simple system which makes multiple, low-granularity changes to the weapon's flight path. Other weapons such as cruise missiles would normally employ a complex system of an autopilot that can fly a route, supplemented by a radar altimeter and possibly a terrain data base to enable covert approaches to the target. Clearly, the level of integration of the weapon's subsystems is a complex subject in its own right, which is not covered in this book.

1.3.6   Stand-Off Range

The final requirement for a guided weapon relates to the range at which it can be launched against its intended target. Ballistic bombs depend solely on the kinetic and potential energy at launch. For low-level bombing, this severely restricts the range the weapon can travel before it impacts the ground. For greater stand-off range, bombing from higher altitudes can be used but then the flight path of the unguided ballistic bomb becomes less predictable as aerodynamic forces and weather influences such as air density, wind speed and direction start to have an effect.

However, greater stand-off range enables the launch aircraft to remain out of range of air defences, thereby improving survivability. Weapon stand-off range can be extended by a number of means, the simplest being to add aerodynamic surfaces (e.g. the addition of a deployable wing kit such as that employed by the United States' Small Diameter Bomb II (SDB II – see Figure 1.1) that provide lift, thereby reducing the decent rate due to gravity).

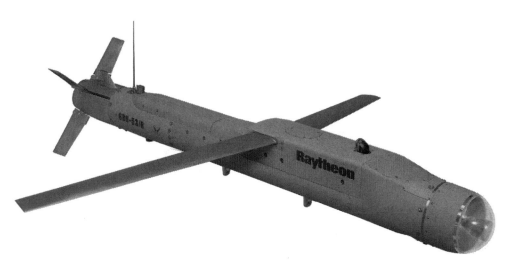

Figure 1.1 Small Diameter Bomb II. Reproduced by permission of Raytheon Missile Systems

For a greater stand-off range, aerodynamic lifting surfaces must be supplemented by a power source to provide thrust. Rocket motors can give high acceleration and boost the weapon to very high speeds (>> Mach 1), albeit with a relatively short burn time. This is very useful for weapons such as Air-to-Air missiles where time to complete the engagement is critical. For even greater stand-off ranges, turbojets (e.g. in a cruise missile) or a ramjets (for a long range, high-speed engagement) are also employed.

1.3.7 Typical Weapon Configurations

From the brief overview of the requirements that drive weapon design, it is clear that a typical guided weapon contains a number of individual subsystems that all contribute to a successful target engagement. These subsystems require a power source, and in missiles with a relatively short time of flight, thermal batteries are typically employed (a thermal battery is a chemical device initiated by a pyrotechnic squib; the battery electrolyte, which is normally a solid and relatively inert, melts and the battery becomes active). For longer range weapons such as cruise missiles, internal power may be provided by a generator turned by a power take-off shaft from the turbojet engine. Weapons may also contain a post-launch data link which can be used to receive target position updates.

The following figures show three typical weapon configurations, covering a Precision-Guided Bomb (Figure 1.2), an Air-to-Air missile (Figure 1.3) and a cruise missile (Figure 1.4).

1.3.8 Implications for the Launch Aircraft

The primary roles of the launch aircraft from the weapon's viewpoint are to transport it to a point where it can be released against its target, to ensure that it is operational prior to launch, to provide the necessary priming to ensure that it has the best chance of engaging the target,

Figure 1.2 Paveway™ IV Precision-Guided Bomb. Reproduced by permission of Raytheon Missile Systems

Figure 1.3   Meteor Air-to-Air missile. Reproduced by permission of MBDA / Thierry Wurtz

and to ensure that it separates cleanly from the aircraft when launched. The weapon cannot perform its task unless it has been correctly initialised by the aircraft. For example, the aircraft will have to provide power to the weapon pre-launch for the purposes of target programming, testing (by built-in test functionality within the weapon) and aligning the weapon's navigation subsystem and sensors. When powered, the weapon may also provide

Figure 1.4 Storm shadow cruise missile on fuselage stations of Tornado GR4. Reproduced by permission of Geoffrey Lee, Planefocus Limited

data on request by the aircraft which includes its type and its operational status and data for reading back stored target data.

From a systems integration viewpoint, it is the electrical connector that provides the primary interface. It is through this interface that the aircraft ensures that the weapon is operational prior to launch and provides the necessary priming to ensure that the weapon has the required precision for a successful target engagement.

The weapon will have a number of other interfaces with the launch aircraft which include physical interfaces such as the mechanical attachments, fuzing lanyards and the electrical connector. However, other interfaces presented by the environment (e.g. aerodynamic forces, physical forces such as vibration and the electromagnetic environment) also exist. The mechanical interface will be covered in Section 1.4 with other interfaces covered in Chapters 9 and 10.

Considering the primary weapon requirements, Table 1.3 identifies the launch aircraft's contribution to these.

These aspects will be discussed in more detail in later chapters.

Table 1.3 Launch aircraft contribution to satisfying primary weapon requirements

Weapon requirement	Aircraft contribution
Lethality	If the weapon has a programmable fuze and a guidance system, then the aircraft would set the relevant parameters to optimise the performance of the warhead against the target. This could include delays, burst heights and required impact attitude
Precision	If the weapon has a guidance system to achieve its engagement precision, the aircraft needs to ensure that the weapon knows its location pre-launch by initialising and aligning the weapon's navigation subsystem to the aircraft's own systems. These could be inertial systems and/or a GPS receiver. For a weapon to initialise its GPS receiver, aircraft data and possibly a precision time input may be required. The aircraft may also need to identify the target's actual location or for some weapons, a flight path to the target
Stand-off range	As range is a function of the weapon itself (i.e. is it a ballistic or a glide weapon or does it have a rocket motor or turbojet), apart from release height, there is little that the aircraft can do to directly affect this. However, it will be incumbent on the aircraft to ensure that the weapon releases safely and avoids collisions with other weapons which may just have been released but which may still be in close proximity to the aircraft

1.4   Carriage Systems

1.4.1   Mechanical Attachments

The advent of smart weapons has driven the need to have smart multi-weapon carriage systems which are capable of controlling modern guided weapons. This section will consider the basic carriage systems employed and then discuss how these have been adopted and, in some cases, coupled with complex electronics to realise a smart multi-weapon carrier.

The deployment of air-launched weapons can be achieved in two ways: downward ejection or, for weapons with a rocket motor, forward firing.

1.4.2   Downward Ejection

NATO aircraft and downward ejected stores generally use common 'hook and eye' mechanical attachments defined by the NATO Standardisation Agreement STANAG 3726. The hook mechanism is part of a pneumatic release unit or bomb rack operated either by the initiation of a pyrotechnic cartridge (the product of which is a high pressure hot gas) or by the release of compressed (cold) gas. In older systems or systems that are used on aircraft with a relatively low speed, bomb racks may be operated by energising a solenoid that opens the hooks, allowing the weapon to fall away under the effect of gravity.

In pneumatic systems, in addition to opening the hooks, the gas is also used to operate rams that push the weapon away from the aircraft to aid aerodynamic separation. Figure 1.5 shows a typical bomb rack.

After the weapon has been mechanically attached to the aircraft (by closing the hooks to grab the eyes fitted to the weapon (bale lugs)), the fuzing attachments are made. These usually consist of mechanical lanyards which remove locking pins from the fuzing mechanism during

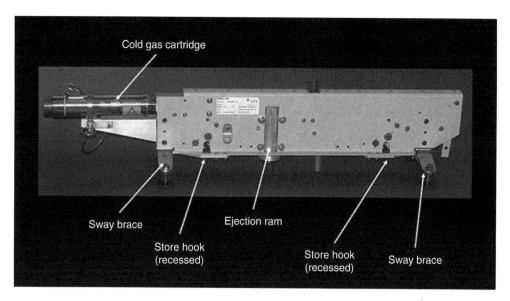

Cold gas cartridge

Sway brace

Ejection ram

Store hook
(recessed)

Store hook
(recessed)

Sway brace

Figure 1.5   Typical bomb rack

the bomb release sequence. However, modern programmable fuzes may contain a thermal battery that needs to be activated at the time of release. In this case, a fuzing power supply is switched through to the fuze by the aircraft on release.

Other aspects of the physical interfacing of weapons such as connector locations in relation to the mechanical attachment points are defined by MIL-STD-8591.

Some Air-to-Air missiles can also be downward ejected. However, for an eject-launched missile, there are currently no standardised mechanical interfaces, and it is left to the WDO to devise the optimum mechanical interfaces.

1.4.3   Forward Firing

Air-to-Air missiles and some Air-to-Ground weapons can be launched by igniting the rocket motor and firing the weapon along a rail. For a rail-launched missile, the mechanical attachments of the launcher are defined by STANAG 3842AA. However, a shortfall of the standard is that there are few details of the mechanical attachments which the weapon must use to attach to the launcher rail. The WDO is therefore required to design missile hangers which interface correctly with the defined rails while sustaining the loads for the carriage and operation of the weapon. The rail is usually a component in a launcher which itself may include electronics to convert power and signals from the launch aircraft to the power supplies and signals required by the missile. A missile launcher could therefore be a typical example of single station smart weapon carriage system.

1.4.4   Multi-weapon Carriage Systems

The earliest multi-weapon carriage systems were used to increase the carriage capability (weapon load-out) of ballistic bombs. These types of carriers provided an extension to the main aircraft carriage system, duplicating the bomb rack and the fuzing connections for multiple bombs. The advent of smart weapons has led to electronics being included in the carrier to convert the interface with a single aircraft electrical connector to multiple interfaces, one for each weapon.

Whilst a smart multi-weapon carriage system can increase the overall weapon load-out, this has implications for the aircraft systems integration activity. The aircraft now communicates with the carrier and not necessarily directly with each individual weapon on the carrier. The aircraft will therefore need to know exactly what is loaded on the carrier and at which station so that, for example, the correct navigation system offsets can be applied and target and fuzing data can be passed to the right weapon. Such a configuration will also increase the system data transmission delays, which could mean that the data received by the weapon is stale. In turn, this could have implications for the accuracy of the weapon when launched against a target. This will be discussed further in Chapter 9.

Further Reading

Fleeman, E.L. (2001) *Tactical Missile Design*, American Institute of Aeronautics and Astronautics Incorporated, Reston.

NATO Standardisation Agency. (1997) Bail (Portal) Lugs for the Suspension of Aircraft Stores. *STANAG 3726*, NATO Standardisation Agency, Brussels.

NATO Standardisation Agency. (2000) Rail Launched Missile/Launcher Mechanical Interface. *STANAG 3842AA*, NATO Standardisation Agency, Brussels.

Rouse, J.F. (2000) *Brassey's Land Warfare into the 21st Century, vol. 8: Guided Weapons*, 4th edn., Brassey's UK, London.

United States Department of Defense. (2010) Standard Airborne Stores, Suspension Equipment and Aircraft – Store Interface (Carriage Phase). *Military Standard 8591*, US Department of Defense, Philadelphia.

2

An Introduction to the Integration Process

2.1 Chapter Summary

This chapter will outline the basic systems engineering process employed in a typical weapons integration programme, from Requirements Definition and the partitioning of requirements to subsystems, through systems implementation to qualification and certification. The need to consider safety from the outset and its place in the overall systems integration process will also be covered as will the individual responsibilities of the aircraft and weapon contractors. The various aspects of the systems engineering process will then be discussed in more detail starting with Requirements Capture.

2.2 Introduction

The integration of a weapon with an aircraft poses a complex systems engineering problem. 'Systems engineering' is an often used term that is frequently misunderstood, largely because it relates to a contextual, bounded set of components, the total of which is greater than the sum of the parts.

Systems engineering has its roots in the American space and defence industry, where it was identified that there was a need to ensure that complex systems performed as required, could be delivered in a viable timeframe and at an affordable cost. The Internet provides many definitions of 'systems engineering', but the most authoritative definition is probably from the International Council on Systems Engineering (INCOSE), which defines it as 'An interdisciplinary approach and means to enable the realization of successful systems' (INCOSE, 2000, Section 2.3). Systems engineering is about applying a set of processes that can be used in a logical and systematic way to solve complex engineering problems. The Defense Acquisition

Aircraft Systems Integration of Air-Launched Weapons, First Edition. Keith A. Rigby.
© 2013 John Wiley & Sons, Ltd. Published 2013 by John Wiley & Sons, Ltd.

University (a US military training establishment that trains Department of Defense (DoD) personnel) has identified the systems engineering process as a technical management and problem-solving process applied through all stages of development, to transform needs and requirements into a set of system product and process descriptions (adding value and detail with each level of development).

So what is systems engineering in the context of weapons integration? From the INCOSE definition, the 'successful system' would be a combination of an aircraft and a weapon that successfully operate together to provide a military effect. To achieve this, the systems engineering process is applied through a number of stages of development to devolve customer requirements into a set of subsystems that result in the end aircraft/weapon system product. At each level of decomposition of the requirements, greater detail and required behaviour of the subsystems are developed, partitioning functions to system hardware and software. Once the decomposed requirements have been implemented in hardware and/or software, then there is a need to prove that the implemented system actually satisfies the high-level customer requirements. In doing so, iteration of the derived requirements, the implementation and the proving are likely to occur. The overall process today is often represented pictorially in what is known, due to its layout, as the V-Diagram.

2.3   The V-Diagram

The typical V-Diagram in its simplest form is depicted in Figure 2.1. The technical process flow proceeds from the upper left-hand side of the diagram, starting with a definition of the requirements, with successive decomposition of those requirements into more detailed levels that can ultimately be implemented into the elements of a system design. The individual elements of hardware and software are then progressively integrated into subsystems and systems, with the proving of components until the completed system is realised at the upper right-hand side of the diagram. On this right-hand leg of the V, the implementation is verified and validated with the corresponding requirements of the requirements decomposition process.

These technical processes are supported by a number of technical management processes, which provide analysis and decision-making tools for configuration control, risk management and the management of interfaces. Whilst this is very much the idealised systems engineering process model, it is possible to expand the V-Diagram to show aspects and sub-processes that are specific to a weapons integration programme and provide examples of the decomposition, implementation and subsequent validation and verification activities.

2.4   Responsibilities

Essentially, an aircraft/weapons integration programme is conducted by the organisations responsible for the aircraft and the weapon. However, for a successful programme, it is important to recognise that there are actually many stakeholders, and the needs of each must be addressed (Figure 2.2 identifies the typical stakeholders in a programme).

During the Requirements Capture phase, it will be essential to understand and capture the requirements from the various stakeholders. These will include military need (for the System User), certification methodologies (for the Certification Organisation) and even the programme cost (a consideration for the nation's tax-payers). However, in actually discharging the

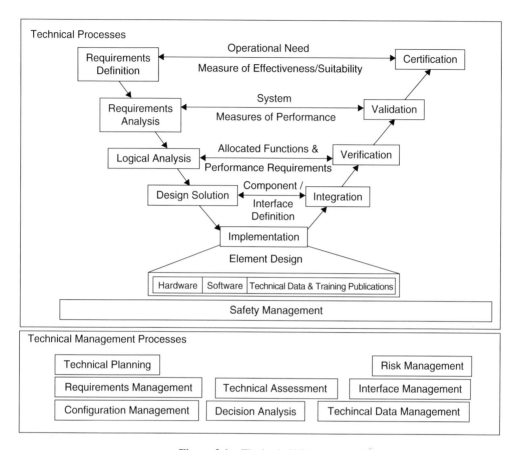

Figure 2.1 The basic V-Diagram

programme, the key stakeholders are the Aircraft Design Organisation (ADO), the WDO and the Contracting Agency. The Contracting Agency (essentially, the programme's customer) will be responsible for capturing the majority of the stakeholder requirements and feeding these into the integration programme. Whilst at this stage, ideally there should be trading of requirements in order to manage programme risk, it is essential that the other key stakeholders (the Aircraft and Weapon Design Organisations (DOs)) are included so that a manageable and practical set of requirements are compiled such that the risk and therefore the cost and time-scales associated with the overall programme can be managed. For example, for a GPS-guided weapon, a requirement for the launch aircraft to be able to attack targets of opportunity may be impractical if the aircraft does not have a means of geo-locating the target with sufficient accuracy to enable a precision engagement.

The responsibilities of the Aircraft and Weapon DOs are to ensure that the integration requirements agreed with the Contracting Agency are implemented and proven in line with the programme's costs and timescales. For this to be successful, there is a need to develop a close working relationship between the two organisations. However, agreeing contractual terms which enable an environment to be developed where such a relationship can flourish is in itself a difficult task and is beyond the scope of this book. The relationship must have trust between

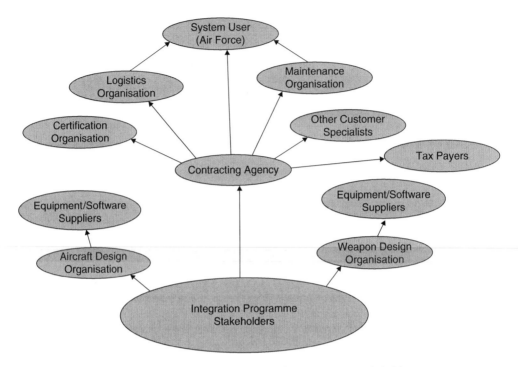

Figure 2.2   Aircraft/weapons integration programme stakeholders

the parties and develop an attitude to solving integration problems which is mutually beneficial for all stakeholders. And this clearly means that there is a need for a close tri-partite relationship between the two organisations and the Contracting Agency.

2.5 Safety

In any systems engineering activity, the need to consider safety at the outset is paramount (in the V-Diagram in Figure 2.1, safety is shown as an all-pervasive technical activity that exists throughout the life cycle of the system). The required safety properties of a system cannot be added after the system has been implemented without recourse to programme cost and time-scale escalation. Therefore, safety requirements must be considered from the outset as part of the requirements set.

The International Civil Aviation Organisation (ICAO) Safety Management Manual (document 9859) defines safety as 'the state in which the risk of harm to persons or property damage is reduced to, and maintained at or below, an acceptable level through a continuing process of hazard identification and risk management'. The management of safety starts with identifying the hazards, where a hazard is defined as 'any real or potential condition that can cause injury, illness or death to personnel; damage to or loss of a system, equipment or property, or damage to the environment' (reference MIL-STD-882). A hazard, in itself, does not cause harm unless there is also a causal element. For example, it may be considered that releasing a weapon from an aircraft could be hazardous. However, an unsafe condition does

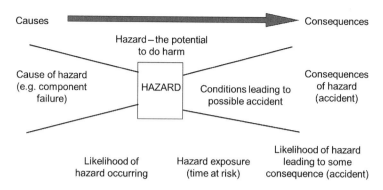

Figure 2.3   The 'Bow-tie' Model

not occur unless there is some further contributing factor such as releasing the weapon when the aircraft is in an attitude which causes the weapon to collide with the aircraft during separation. Therefore, a hazard has the potential to cause harm (i.e. there is a consequence) but only if there is a further contributing factor. Also a hazard may not always be present. In this example, the hazard does not exist until the weapon is released. Therefore, in understanding safety, there is the likelihood that the hazard will occur, a period of time when the hazard is present, and the likelihood that the hazard will lead to a consequence (an accident). The causes/hazard/consequence model (the 'Bow-tie' Model) is shown in Figure 2.3.

Safety is considered to be a non-functional requirement in that it is a property of the system rather than an intrinsic part of its functionality. In an ideal world, the system would be designed such that the consequences would never occur. However, this is impracticable in a real system and the risk of the consequence occurring will, in reality, always be present. Therefore, system safety requirements must be balanced with other program requirements such as performance and cost. In order to do this, it is essential to identify the hazards and to understand the severity of the consequences and the likelihood of them occurring throughout the life of the system. This then enables the consequences to be guarded against without having an undue negative effect on system performance or the cost of the weapons integration programme.

There are a number of methods which can be employed to analyse the severity of consequences, but the most common is to undertake a Functional Failure Analysis (FFA). Such an analysis will relate the failure of the main system functions and any contributing factors to the hazard that they create. In the example (releasing a weapon when the aircraft is in an unsafe attitude), the FFA would identify that a failure of the weapon release function when the aircraft is in an unsafe attitude could have catastrophic consequences. The contributing factor in this case is the aircraft being in an unsafe attitude for the safe release of the weapon.

For military systems, the prevalent standard governing the management of safety is MIL-STD-882 published by the United States DoD. This standard provides a process for relating the severity of a consequence, with its probability of occurrence and, from this, identifying a Hazard Risk Index (HRI). A common methodology used to analyse the effects of failures within a system is Fault Tree Analysis (FTA), a methodology originally developed in the early 1960s to support safety analysis of military systems. FTA determines, in a logical way, which failure modes at one level produce critical failures at a higher level in the system. The methodology is defined in more detail in MIL-HDBK-338.

It is the HRI which enables the weapons integrator to determine if action needs to be taken to reduce any residual risks. Tables 2.1, 2.2 and 2.3 include example definitions of Hazard Severity Categories, Probability Categories and HRI.

Table 2.1 Hazard severity categories

Severity	Consequence
Catastrophic	Could result in death, permanent total disability, total system loss or irreversible severe environmental damage
Critical	Could result in severe injury or severe occupational illness, major system damage or reversible environmental damage
Marginal	Could result in major injury or minor occupational illness, minor system damage or minor environmental damage
Negligible	Could result in less than minor injury or occupational illness, less than minor system damage or minimal environmental damage

Table 2.2 Hazard probability categories

Probability	Frequency of occurrence
Frequent	Continuously experienced. For example, could occur greater than once every 10 h
Probable	Will occur frequently during the life of the system. For example, could occur greater than once every 100 h
Occasional	Will occur several times during the life of the system. For example, could occur greater than once every 1000 h
Remote	Unlikely, but can reasonably be expected to occur during the life of the system. For example, could occur greater than once every 1 million hours
Improbable	Unlikely to occur during the life of the system, but possible. For example, could occur less than once every 1 million hours

Table 2.3 Hazard Risk Index

		Severity			
		Catastrophic	Critical	Marginal	Negligible
Probability	Frequent	1	3	7	13
	Probable	2	5	9	16
	Occasional	4	6	11	18
	Remote	8	10	14	19
	Improbable	12	15	17	20

HRI	
1–5	High hazard risk – unacceptable
6–11	Moderate hazard risk – may be acceptable with review.
12–20	Low hazard risk – acceptable with less justification.

Table 2.4 FAA safety order of precedence

Description	Priority	Definition
Design for minimum risk	1	Design to eliminate risks. If the identified risk cannot be eliminated, reduce it to an acceptable level through design selection
Incorporate safety devices	2	If identified risks cannot be eliminated through design selection, reduce the risk via the use of fixed, automatic or other safety design features or devices. Provisions shall be made for periodic functional checks of safety devices
Provide warning devices	3	When neither design nor safety devices can effectively eliminate identified risks or adequately reduce risk, devices shall be used to detect the condition and to produce an adequate warning signal. Warning signals and their application shall be designed to minimise the likelihood of inappropriate human reaction and response. Warning signs and placards shall be provided to alert operational and support personnel of such risks as exposure to high voltage and heavy objects
Develop procedures and training	4	Where it is impractical to eliminate risks through design selection or specific safety and warning devices, procedures and training are used. However, concurrence of authority is usually required when procedures and training are applied to reduce risks of catastrophic, hazardous, major or critical severity

Source: FAA (2000)

From the safety analysis, it is possible to determine what course of action should be taken. For the highest hazard risk (HRI ≤5), action must be taken (e.g. a change to the design) in order to improve the HRI and therefore ensuring that safety can be maintained.

However, as noted earlier, to have a completely safe system design may be impracticable, if not impossible. There could therefore be a need to rely on other factors such as safety devices, warning devices or procedures/personnel training. The Federal Aviation Administration (FAA) articulates this as reproduced in Table 2.4.

Whilst much of this safety discussion relates to the probability of an event occurring, it should be noted that the design of a weapon should be such that time-related failures (e.g. a structural failure of the weapon) are likely to have a low probability of occurring. The most significant weapon hazards in relation to the aircraft are likely to be present during key events such as missile firing (e.g. a rocket motor may be relatively inert during air carriage of the missile, but when it is initiated, its very function could be hazardous as the fuel burns and pressure builds inside the rocket motor casing). It is the risk of the consequence occurring during an event that usually poses the greatest risk to the aircraft.

For a weapons integration programme, it is necessary to understand the lifetime risk to the aircraft/weapon combination (normally addressed in terms of the risk of a consequence occurring per hour of operation) to ensure that the system implementation is safe, but also the event-based risk needs to be considered. Whilst this will largely be a requirement on the design of the weapon itself, cost and performance will also be a factor in determining the levels of safety implemented in the design. However, it is not unusual for key events which could be catastrophic to have a large safety margin added during design, thereby reducing the risk to the launch aircraft.

This section has given a brief introduction to safety. However, the subject will be revisited in Chapter 9, where an aircraft/weapons integration scenario is considered.

2.6   The Use of Requirements Management Tools in the Systems Engineering Process

From requirements definition to certification, a weapons integration programme will create large amounts of data. Keeping track of the flow-down of requirements on the left-hand side of the V-Diagram to the progressive build-up of a certified system on the right-hand side, there is a need to track the development of lower level requirements and how the design is then verified against these. To efficiently manage this process, large programmes invest in complex tools which are capable of capturing the top-level requirements and decomposing these down to the lower levels of design and then capturing the results of testing and analysis, thereby providing a through-programme picture that all requirements have been satisfied. Such tools can bring great benefits to the control of the programme and provide a central repository for all relevant data that is accessible by all the people involved in the integration programme. Whilst it is not uncommon for large aerospace organisations to employ such tools, for very simple integration programmes, the author has used spreadsheets to great effect. In order to remove the dependence on any one requirements management tool, later chapters will employ a tabular method. However, this comes with a warning. As the complexity of the integration programme increases, a spreadsheet soon becomes unmanageable, and it is then that a complex requirements management tool becomes essential.

2.7   Weapons Integration Requirements Capture

Having introduced the V-Diagram, we will first concentrate on parts of the V-Diagram which relate to the capture and management of the top-level requirements for the integration programme. The remaining parts of the diagram will be covered in Chapter 3. Figure 2.4 shows the key activities in defining the top-level requirements.

As noted earlier, all stakeholder needs must be identified. Primarily, these will be defined by the Contracting Agency, but other needs must be identified and acknowledged from the start. For example, if the Contracting Agency requires an existing developed weapon to be integrated with the aircraft, then identifying requirements that invalidate the existing design certification of the weapon may cause problems for the WDO when they are required to provide qualification evidence. This could result in a need to re-test, or in the worst case, re-design the weapon, where there would be cost and timescale implications.

Stakeholder requirements will fall into a number of categories which may include system performance criteria; operational requirements/use cases (i.e. a scenario which defines the intended use of the system, for example, the need for the system to be able to release a weapon); personnel skill/training needs; applicable national and international standards covering, for example, safety, security, quality and electromagnetic compatibility; human factors; reliability; supportability; and interoperability. It is therefore essential to capture all the requirements of the various stakeholders (the Contracting Agency's stakeholders are shown in Figure 2.2) although it is also likely that some of the requirements may conflict.

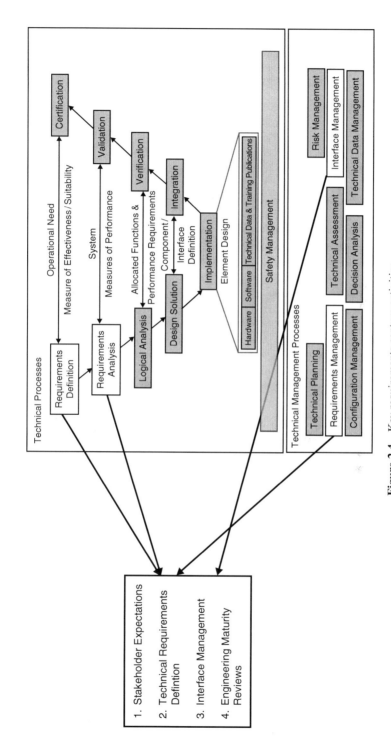

Figure 2.4 Key requirements capture activities

Therefore, in capturing requirements, there is a need to understand those which are essential and those which may be desirable. Having captured the stakeholder requirements, these must be transferred into unique, quantitative and measurable technical requirements expressed as 'shall' statements that will be used in defining a design solution. In defining these requirements, trade studies may be undertaken and concepts explored and modelled in order to gain a better understanding of the adequacy of the requirements and their feasibility and to identify any emergent properties of interacting or conflicting requirements. The requirements may also be categorised into 'Key', 'Important' and 'Routine' requirements so that an order of precedence can be determined. This will help when requirements need to be traded. The 'Key' requirements are the most important and, wherever practicable, must be achieved, possibly at the expense of less important requirements.

At this point, we must not lose sight of the fact that an aircraft/weapons integration programme requires two, often separately designed, systems to be brought together. The interfaces between the two systems must therefore be managed from the outset. An interface management process will be required that is used by both the Aircraft and Weapon DOs. Also to ease documenting of the interface, a format for an ICD will be required. This will be discussed further in Chapter 8.

Therefore, before Requirements Capture commences, the programme should have robust processes in place for Technical Planning, Technical Risk Management, Configuration Management and Technical Data Management. It is essential that each transition around the V-Diagram is accompanied by a review of the level of maturity that has been achieved. Such management reviews exist to review the engineering outputs from the previous phase to ensure that they are adequate to form the inputs to the subsequent phase. MIL-STD-1521B (which has been withdrawn by the DoD but still identifies good practice) identifies a number of such reviews that are used throughout the life cycle of a programme. At the end of the Requirements Capture phase, the corresponding review would be the System Requirements Review (SRR). It is the SRR which confirms that a mature set of requirements exists, sufficient cognisance of the stakeholder needs has been taken and that the management processes are in place that will lead to a successful programme.

2.8   The Need for Unambiguous, Clear and Appropriate Requirements

There is a danger when defining requirements that emphasis is placed on the more obvious stakeholder needs to the detriment of less interesting needs. During a weapons integration programme, the boundary of the problem (usually largely aircraft-centric with only the aspects of the weapon directly relating to integration being captured) must be understood. To enable a check to be made that all areas have been considered and that all the requirements have been captured, a Work Breakdown Structure (WBS) must be defined. An example WBS is depicted in Figure 2.5. In this figure, the WBS activities that relate specifically to the systems integration aspects of weapons integration are shown in white boxes. Activities relating to other aspects of weapons integration are shown in grey boxes.

When defining the system requirements, it is likely that there will be many which are in conflict, particularly when the aircraft requirements are compared with the weapon requirements. This can lead to ambiguity. Table 2.5 shows conflicting aircraft and weapon requirements.

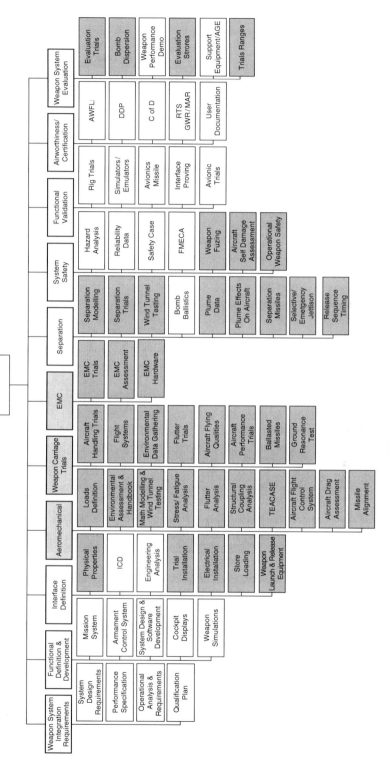

Figure 2.5 Example WBS

Table 2.5 Reconciling an initial requirement into an agreed requirement

Requirement number	Initial requirement	Aircraft constraint	Weapon constraint	Agreed requirement
00100	The weapon shall engage the target with a CEP of 3 m	Maximum update rate of data parameters is 50 Hz	To meet CEP, all position and velocity data parameters must be updated at 100 Hz	The weapon shall engage the target with a CEP of 6 m

Note: CEP is 'Circular Error Probability' – a measure of the accuracy with which the weapon hits its target (see Chapter 5).

The conflict introduced by the constraints is such that neither of the DOs could accept the initial requirement. Therefore, there would be a need to understand the implications for both the aircraft and the weapon designs if this requirement was to be maintained. From an aircraft viewpoint, there could be a need to re-design the parts of the system which generate the data parameters in question. For an existing aircraft, this is likely to be a significant change which may drive the overall programme to be unaffordable for the customer. Similarly, for the weapon, to accept parameters which have a level of staleness may require changes to its navigation system or the introduction of complex software filtering and prediction methods to overcome the shortfalls of the aircraft system. In this example, it is therefore necessary to investigate all the options, undertaking modelling and trade studies and then present the options and their consequences to the Contracting Agency for resolution. The outcome could be that an expensive and time-consuming modification to either the aircraft or weapon is acceptable. However, from the results of modelling, it may be found that with the aircraft providing data updates at 50 Hz, the overall Circular Error Probability (CEP) of the weapon is not greatly impacted (as shown as the Agreed Requirement in Table 2.5). Therefore, in undertaking such an analysis, the two DOs in conjunction with the Contracting Agency have reduced integration risk at the outset and found a resulting unambiguous requirement which is also acceptable to the Contracting Agency. This level of risk mitigation very early in the integration programme avoids a mismatch in constraints becoming a serious integration issue later in the programme where the impact on timescales and costs would be detrimental.

Of course, this is a simple example used to demonstrate a point. In reality, the wider implications of reducing the accuracy of the weapon system may be such that the user's military needs cannot be satisfied. It is therefore essential that any tailoring of individual requirements is agreed by the Contracting Agency with the wider stakeholders.

All requirements should be unambiguous, clear and appropriate (the example certainly satisfies the first two checks). All requirements must be appropriate for the level of definition, that is, the position on the V-Diagram. At this stage, we are trying to capture the high-level needs of the stakeholders and so the example is appropriate. At this level, it would not be expected to see requirements that relate to the specific data parameters being passed between the aircraft and the weapon. Such requirements would be developed through the process of requirements decomposition and partitioning and would be documented in an ICD.

Figure 2.6 shows the V-Diagram modified to capture the Contracting Agency's Requirements Definition (often referred to as the Statement of Requirements or User Requirements) and the

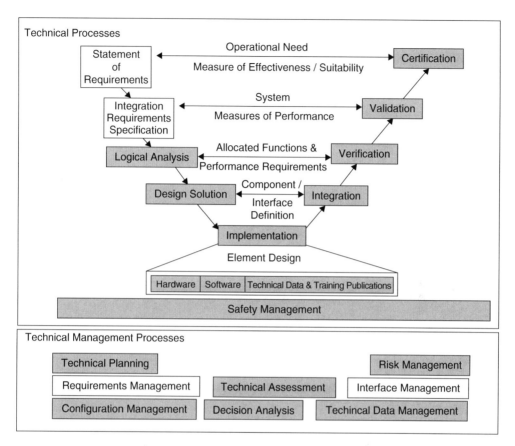

Figure 2.6 Modified V-Diagram

Requirements Analysis that results in a three-way agreement between the Contracting Agency and the two DOs (the Integration Requirements Specification – IRS). It is the IRS which captures all the high-level requirements to which the DOs can be contracted to deliver and so it needs to be agreed early in the programme (ideally, before any integration contracts are placed by the Contracting Agency). The IRS will be covered in more detail in Chapter 8.

2.9 Minimising Requirements

It is likely that every high-level requirement will spawn a number of child requirements which in turn, when further decomposed and partitioned to the system architecture, will generate a number of grandchild requirements. It is not uncommon for a single high-level requirement to result in tens or even hundreds of lower level requirements. For each requirement generated, there will have to be an associated verification or validation activity, and all this must be adequately captured and controlled. For this reason, a typical weapons integration programme will employ computer-based Requirements Management tools to enable capture, tracking and validation/verification to be accurately controlled.

Although it is necessary to capture all the stakeholder needs as requirements, these must be minimised in order to avoid too much detail needing to be managed early in the programme. This will then enable an orderly flow-down of requirements to the lower levels of design, resulting in a manageable set of activities on the right-hand side of the V-Diagram.

Further Reading

Federal Aviation Administration. (2000) Principles of system safety, in *System Safety Handbook*, Federal Aviation Administration, Washington, DC, Table 3.7.

International Council on Systems Engineering (INCOSE). (2000) *Systems Engineering Handbook Version 2.0*, INCOSE, Seattle.

United States Department of Defense. (1985) Military Standard, Technical Reviews and Audits for Systems, Equipments and Computer Software. *Military Standard 1521B*, US Department of Defense, Philadelphia.

United States Department of Defense. (1998) Electronic Reliability Design Handbook. *Military Handbook 338*, US Department of Defense, Philadelphia.

United States Department of Defense. (2000) Department of Defense Standard Practice for System Safety. *Military Standard 882*, US Department of Defense, Philadelphia.

3

Requirements Analysis, Partitioning, Implementation in Aircraft Subsystems

3.1 Chapter Summary

Once the top-level integration requirements have been defined, there is a need to decompose these into more detailed requirements and to then partition them to the aircraft subsystems. This partitioning exercise depends on the actual aircraft system architecture and will therefore differ between aircraft. The requirements partitioned to individual aircraft subsystems will likely be implemented in both hardware and software. Each requirement placed on the aircraft's subsystems will also need to be proven for correct implementation, and the subsystems will then be progressively built up into an overall integrated system that provides the required military capability. This chapter will discuss requirements decomposition and partitioning and will then discuss the right-hand side of the V-Diagram and the steps taken to develop a Safety Case and to have a certified system that can be employed by the military user. This chapter will discuss these aspects of the systems engineering process and the need to avoid 'over engineering' of requirements in order to ease certification and to control integration costs. A final version of the V-Diagram will then be defined, which will be used in later chapters.

3.2 Introduction

Figure 3.1 shows the key activities in the requirements decomposition and partitioning process.

These activities progressively and iteratively decompose the requirements captured at the top of the left-hand side of the V-Diagram into detailed subsystem requirements and then into detailed specifications which can be implemented in hardware and/or software. These steps will be explored in the following sections of this chapter. However, before analysing and decomposing

Aircraft Systems Integration of Air-Launched Weapons, First Edition. Keith A. Rigby.
© 2013 John Wiley & Sons, Ltd. Published 2013 by John Wiley & Sons, Ltd.

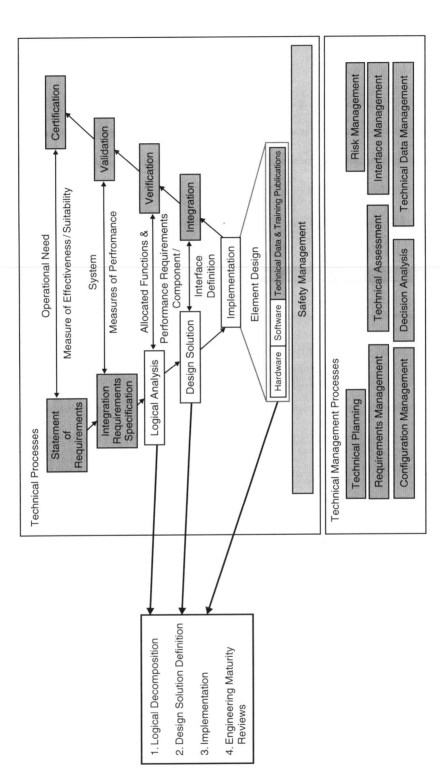

Figure 3.1 Key decomposition and partitioning activities

requirements, there is a need to take an initial view on the underlying system architecture that will enable the Contracting Agency's requirements to be correctly implemented.

3.3   System Architecture

In a real weapons integration programme, it is likely that the aircraft system architecture will already have been defined unless the aircraft programme is also in the early stages of development. There will therefore be an existing system architecture, so to control development and re-certification costs, there will be a need to avoid fundamental changes to this unless the needs of the integration programme cannot be satisfied. An example of this could be a legacy aircraft which does not have a GPS subsystem, but it has been identified that the navigation accuracy which such a subsystem could provide is essential to satisfying the requirements. The introduction of such a subsystem could drive further significant changes to the existing navigation system.

When analysing the required system architecture, there are likely to be several options for the weapons integrator, and these will need to be traded to arrive at an optimum solution which is affordable.

Of course for aircraft which already have a capability to deploy high-performance weapons, an understanding of which subsystems are likely to be affected by a weapons integration programme will already be understood. Figure 3.2 shows the typical manned aircraft system components which contribute to the platform's weapons capability. In this example, the dominant components in the system are the Weapon Aiming Computer and the SMS.

The SMS controls all the management functions for the weapon including priming, arming, release and jettison. For controlling the weapon, there is a need for the crew to have various controls (inputs to the system) and corresponding displays (provided by the Display Computer)

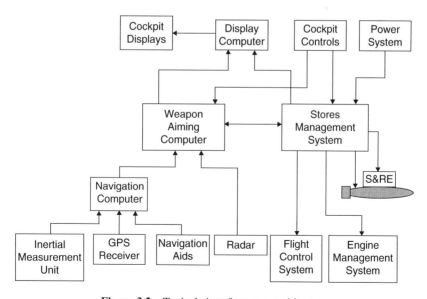

Figure 3.2   Typical aircraft system architecture

for monitoring weapon status and so on. Although all aircraft systems require power, for the armament system, the controlling of power is the key to basic system safety. These aspects and the other interfaces to the Flight Control and Engine Management Systems will be discussed in detail in Chapter 4.

The Weapon Aiming Computer is typically a general-purpose computer that interfaces with a number of other subsystems in order to identify targets and to direct the attack. Clearly, there will be a need for the Weapon Aiming Computer to provide status information and cues to the crew and to respond to selection inputs from the cockpit. As will be seen in Chapter 5, for a successful target engagement, navigation accuracy is important. In this example architecture, a dedicated Navigation Computer is employed to fuse the position, velocity and time elements provided by a number of sensor subsystems (an Inertial Measurement Unit (IMU), a GPS (covered in more detail in Chapter 4) and a range of other navigation aids such as a magnetic compass and radio aids such as Tacan).

The Weapon Aiming Computer also has an interface to the aircraft's radar. For Air-to-Air engagements, this will be used to identify target position and dynamics. In an Air-to-Ground attack role, the radar may be used as an additional sensor to improve positional accuracy of the aircraft with respect to the target.

For a successful detection and engagement of a target, all these subsystems must work together.

3.4 Requirements Decomposition

Having captured the top-level requirements, there is a need to analyse how these will be implemented. However, taking the requirement example from Chapter 2, it is clear that to satisfy the need, a greater level of understanding is required not only of what constitutes the overall system functions that can be derived, but also of the underlying system architecture onto which the requirements will be implemented.

The decomposition process will develop a number of child requirements that introduce an increasing level of detail into the design. In undertaking the decomposition, a greater understanding of the requirements will be gained, and this will necessitate the introduction of other derived functional, temporal or behavioural requirements that will form the input to the partitioning stage where lower level requirements can be allocated to the aircraft's subsystems. For example, the requirement from Chapter 2 may decompose into a number of lower level child requirements as shown in Table 3.1. To support the decomposition process, it may be necessary to model requirements such that derived requirements can be identified (in the table, it is clear that several requirements have needed engineers to model the system in order to understand the target and aircraft positional accuracies and system delays which can be tolerated to achieve the required CEP). In generating these lower level requirements, it can be seen that further new requirements have been derived which, whilst not being explicit in the parent requirement, do contribute to it being satisfied. The comments in the table, whilst not essential, capture a useful commentary on design decisions which may have been taken, so enabling a better understanding of the child requirements should they need to be revisited later in the integration programme.

The set of child requirements is deliberately simplistic to illustrate how decomposition of the top-level requirements could proceed. In reality, there may be many more child requirements, and these would likely start to identify more detailed performance and system behavioural needs. Included in this would be any safety requirements, alluded to in the

Table 3.1 Derived child requirements

Requirement number	Agreed requirement		
00100	The weapon shall engage the target with a CEP of 6 m		

Child requirement number	Requirement	Requirement type	Comments
00101	The aircraft system shall geo-locate the target with an accuracy of 1 m in longitude, latitude and height above mean sea level	Derived Functional	From modelling, to achieve a 6 m CEP, the target location must be known to within 1 m in each direction
00102	The aircraft position in three axes shall be accurate to 1 m in longitude, latitude and height above mean sea level	Derived Functional	From modelling, to achieve a 6 m CEP, the release point of the aircraft must be known to within 1 m in each direction
00103	System processing and data transmission delays from the true release point being reached to a weapon separating from the aircraft shall be less than 25 ms	Implied Temporal	From modelling, to achieve a 6 m CEP, the delays within the system must be guaranteed to be less than 25 ms
00104	Prior to weapon release, an estimate of CEP shall be displayed to the crew	Derived Functional	Provides a check that collateral damage will be minimised
00105	The weapon aiming subsystem shall account for the ballistic flight of the weapon, post release, to ensure that an accurate CEP prediction (3-sigma accuracy of 0.1 m) is calculated	Derived Functional	An estimated CEP to the crew must be accurate if it is to be relied upon

example as new requirements that relate to the protection of innocent parties and the reduction of collateral damage (Child Requirement 00105).

3.5 Requirements Partitioning

In partitioning requirements, there may be a number of possible options. These will need to be studied and possibly modelled in order to arrive at an optimum solution. It is the partitioning activity which drives the design of the underlying architecture of subsystems.

Requirement 00101 from Table 3.1 would typically be mapped on to the Weapon Aiming Computer depicted in Figure 3.2. However, the ultimate accuracy by which the system can geo-locate a target will depend on how accurately the system can geo-locate itself, and for this, other elements of the system such as the accuracy of the data generated by the Navigation Computer (which itself will have a dependence on the accuracy of the sensors feeding it) will be critical. Requirement 00102 recognises this dependency.

Modelling the system would enable the integration engineer to understand the sensors which provide the greatest contribution to providing an accurate aircraft position. For example, it may be determined from modelling that the IMU combined with other available navigation aids can provide an acceptable solution. Equally, the modelling may show that the accuracy may need to be improved by the addition of a precision sensor such as a GPS receiver.

Finally, the radar in a ground-mapping mode could be used to locate the target with the Weapon Aiming Computer fusing all the data inputs into an accurate estimate of the target's actual location.

In this example, the child requirements from Table 3.1 can be allocated to a number of different subsystems, although it should be recognised that simply flowing down a requirement to say, the Navigation Computer may not guarantee that the overall accuracy requirements will be met. Further modelling will be necessary to flow down the actual requirement that will provide an adequate contribution to meeting the top-level Contracting Agency requirements. However, the modelling may determine that the Navigation Computer would need to provide position accuracies of better than 0.5 m in all three axes. This may dictate that a more accurate means of navigation would be required with the associated cost and timescale penalties. Alternatively, there may be a need to revisit the top-level Contracting Agency requirements to seek a relaxation. It is the integration engineer who will be required to define not only the required performance of the aircraft's system, but also the likely impact on the top-level requirements.

Partitioning is likely to generate more detailed requirements for each subsystem, and these will need to be captured in subsystem specifications and flowed down into individual equipment and software specifications, as required.

3.6 Subsystem Implementation

The requirements partitioning process will identify all the detailed requirements which each subsystem must implement. For each subsystem element, there will be a need to identify workable and cost-effective solutions, the result of which will be documented in detailed specifications for equipment (hardware, possibly with embedded software) and software, with separate documentation covering the inter- and intra-subsystem ICD's.

In compiling specifications, non-functional requirements such as environmental requirements, design life, system reliability, maintainability needs and individual safety requirements will also be specified. Safety requirements in particular need to be understood as this is not something that can be built in later. To assist in this, there is a need to undertake a FFA where each subsystem function is analysed to determine the consequences of any failures of the function, taking account of any contributing factors (as discussed in Chapter 2). The criticality of a failure can be determined and design requirements introduced into individual specifications so that the consequences of the failure are mitigated.

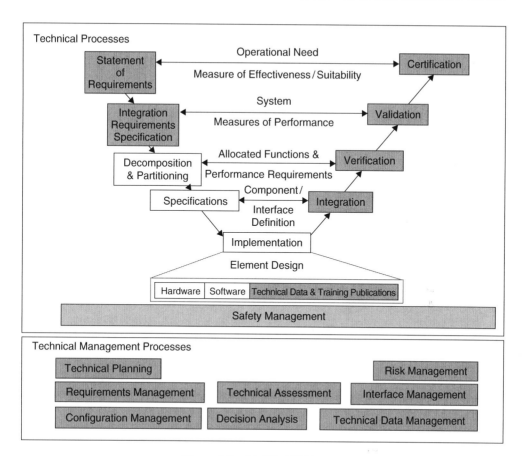

Figure 3.3 Modified V-Diagram

In terms of the V-Diagram, the process has now reached the bottom of the diagram. Here the system products are implemented and tested in line with their specifications. The Logical Analysis and Design Solution parts of the V-Diagram cover the Decomposition, Partitioning and Specification of the system. This is reflected in the revised diagram in Figure 3.3.

3.7 Maturity Reviews

During the course of the decomposition and partitioning processes, project maturity reviews (e.g. as defined by MIL-STD-1521B) must continue to be held to ensure that the design is progressing to plan and that the Contracting Agency's requirements can still be satisfied by the system implementation.

At the aircraft integration level, both Preliminary and Critical Design Reviews should be held. During the implementation phase, it is not uncommon for individual reviews to be undertaken on the lowest levels of implementation such as for individual equipment or software programmes. A Test Readiness Review will also be undertaken to ensure that the individual element is ready to commence testing against its specification.

3.8   Right-Hand Side of the V-Diagram

As noted previously, there will be a need to ensure that every derived requirement has been correctly implemented and that the combination of the functionality provided by the implementation meets the requirements of the Contracting Agency.

For a complex integration programme, there will potentially be thousands of requirements that have been derived and implemented in both hardware and software. In defining requirements, there is therefore a need to also identify the methods which will be used to prove that the requirement has been correctly implemented. For this, it is essential to understand the implications of every requirement which is derived, as every requirement and its associated proving task have an associated cost and timescale. Inappropriate or over engineered requirements will still have to be proven, and this can increase costs and the duration of the programme unnecessarily. It should also be noted that any problems encountered during system proving could drive iteration of lower level requirements which will cause changes to already implemented hardware and software. Therefore, changes to activities that appear on the left-hand side of the V-Diagram resulting from problems identified on the right-hands side of the diagram are very bad for the programme. This is the reason why much thought needs to be applied to defining the proving methods when developing the requirements.

The right-hand side of the V-Diagram covers the progressive proving of the implemented system through integration, verification, validation and ultimately to the Safety Case and a certified system. The specific activities are highlighted in Figure 3.4. Each of these phases of proving will be explored later.

3.9   Proving Methods

There are three primary methods (Test, Analysis and Inspection) that, when used in combination, will prove that an implemented system or subsystem meets its requirements.

Performing a test is the most obvious method of proving that the implemented system meets its requirements. A test will measure the characteristics of the system and will be used to gather specific detailed evidence required for the Safety Case (see Section 3.13) and for eventual certification. Testing will also demonstrate, usually in a representative environment, that the system operates satisfactorily.

Whilst testing can provide a good record of system performance, it is only as good as the number of test cases performed. For example, it is impractical to undertake software testing that covers every single eventuality of inputs, outputs, states and variables (however, for software that is deemed to implement safety-critical functions, the use of Formal Methods – defining the code using verifiable mathematical constructs – may be a preferred solution). Therefore, other methods are required such as Analysis, where compliance to the design is proven through a technical evaluation of the design documentation, and Inspection, where a visual examination or other non-destructive investigation is undertaken to ensure that the physical implementation is as required (e.g. the correct connector part number has been adopted).

As noted earlier, when defining requirements, it is necessary to also define the methods by which implementation will be proven. Table 3.2 shows an example of possible low-level weapons integration requirements with the table expanded to identify the proving methods that will be employed.

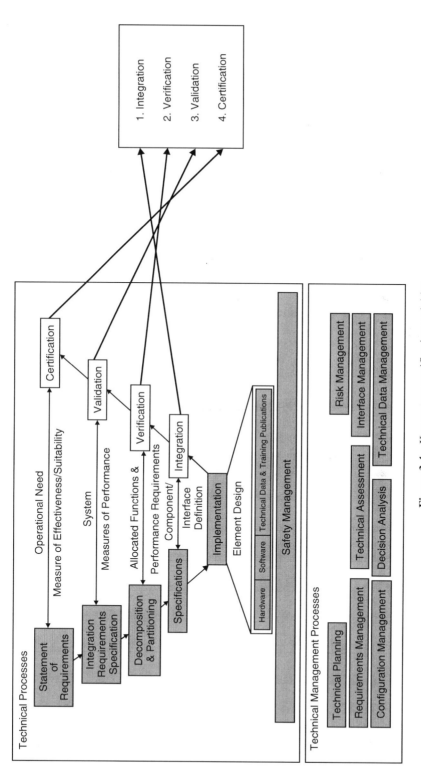

Figure 3.4 Key system certification activities

Table 3.2 Proving methods

Requirement number	Agreed requirement	Child requirement number	Child requirement		Responsibility
00100	The weapon shall engage the target with a CEP of 6 m	00103	System delays from the true release point to a weapon separating from the aircraft shall be less than 25 ms		Aircraft Design Organisation (Mission Systems Team)
Low-level requirement	Requirement	Method type	Outline of activity		
00103.001	The Weapon Aiming Computer shall schedule all acyclic data bus transactions within 2 ms (±10%)	Analysis	Analyse the software and computing architectures of the Weapon Aiming Computer to calculate the worst-case delay to scheduling an acyclic data bus transaction. In undertaking the analysis, it should be assumed that no re-tries are required due to failed communications		
00103.002	The SMS shall respond to a valid request for weapons release within 10 ms (±10%)	Analysis	Analyse the hardware interfaces, software and computing architectures of the SMS to calculate the worst-case delay to generate a weapon release output discrete signal		Aircraft Design Organisation (Mission Systems Team)
00103.003	The operating time of the weapon ejection system shall beless than 10 ms (±10%) from receipt of a valid initiation signal	Test	Undertake a test of the ejection system under all specified environmental conditions to determine the worst-case operating time		Aircraft Design Organisation/Ejection System Supplier

Table 3.2, in a spreadsheet format, could be expanded to include a more detailed description of the proving task, the evidence that will be acquired and a statement on compliancy against each requirement. This would then enable the spreadsheet to be progressively populated throughout the programme as evidence becomes available. The spreadsheet would act as an aid to developing the system Safety Case and therefore benefit the process required to achieve certification.

3.10 Integration

The integration process is used to transform the implemented components of the design into the desired end products through assembly and integration of lower level components. Each system component will have undergone its own proving programme to ensure that it meets its specification. This is likely to include software testing, hardware analysis and testing (including a level of environmental testing to ensure the hardware functions, e.g. across its full specified temperature range) and a level of integration testing to ensure that the component software and hardware operates as required. Each component, when proven, will be delivered to the systems integrator ready for integration with other system components.

During the integration programme, different components of the subsystems will need to be combined to form the overall system. At the lowest level, this could be ensuring that a software programme runs correctly and provides its specified performance for the subsystem element (e.g. a Weapon Aiming Computer). As the subsystems are progressively assembled and tested, then the integration activity would address larger parts of the system, for example, checking that two computers (e.g. a Weapon Aiming Computer and an SMS) communicate correctly, transferring the right data at the right time with the required update rate and latency.

Integration is normally conducted in a SIL which represents the parts of the aircraft system which contribute to the control of weapons and would normally have the ability to employ weapon hardware (in the form of simulators or real hardware in the loop) to aid the process. It is the SIL where much of the test evidence that will be used to certify the system against the Contracting Agency's requirements will be generated. The SIL is also where the aircraft system and the weapon are brought together for the first time and where every aspect of the systems elements of the aircraft/store ICD is tested. Key to this testing will be to ensure that the correct data is generated by the aircraft and passed to the weapon when required, and that the weapon generates the correct responses and that these are actioned correctly by the aircraft. The SIL also enables tolerances to system degradation and faults to be assessed thereby ensuring that the boundary between the combined system working and not working is understood. It is often the stretching test cases that uncover emergent unwanted properties of the implemented system, which in turn may require equipment, subsystem or system requirements to be modified. It is probably impossible to avoid this although the experience of the integration engineers can greatly reduce the number of occurrences.

Once sufficient testing has been performed in the SIL and the evidence analysed, it is likely that the system will be cleared for flight on an instrumented aircraft such that testing can be undertaken in the real system with the aircrew in the loop. Initially, the flight testing will be re-proving the systems aspects tested in the SIL and ensuring that there are no further emergent properties resulting from the dynamics of a flight test programme. It is also during flight testing that system development activities and aeromechanical proving activities come together in the real system environment. Invariably, this will identify further problems which will need to be corrected by an iteration of the systems design.

The relevant documentation produced during integration will present matrices which identify compliance with the relevant specifications.

3.11   Verification

Integration and the resulting iterations of the implemented system should provide an integrated aircraft/weapon capability proven against the lowest level requirements. Whilst the system may function, there is a need to investigate the actual performance that has been achieved and verify that the major system functions provide the required capability. As this is an activity on the right-hand side of the V-Diagram, then verification is an exercise to gather data which supports proving that the higher level performance requirements have been correctly implemented. Where integration relied largely on tests to generate evidence, during verification, it is likely that more analysis techniques will be used to support the practical evidence from integration testing. This would be particularly applicable to safety critical software where it is not practical to completely test for every eventuality. The data produced during verification will give confidence that under extremes of system operation which may not be testable, the system will continue to meet its performance requirements. The data will be captured in a Statement of Design (SoD) document.

3.12   Validation

The right-hand side of the V-Diagram exists to generate data that can ultimately be used to provide a closure of the integration programme and to deliver a working capability to the end user. Validation encompasses the final proving at the system level to ensure that the overall performance of the system meets its requirements. From a purely weapons integration viewpoint, there is a need to ensure that all requirements captured in the IRS (introduced in Section 2.8) have been correctly implemented and that evidence exists to support this (including any evidence which may identify exceptions to the IRS requirements). For every IRS requirement, there should be a comprehensive set of evidence. For example, if there is an IRS requirement which states that the aircraft shall control the weapon via an interface compliant to MIL-STD-1760, then the evidence is likely to consist of a large data package. This could include, for example, documentation proving the correct use of the signal set, the results of an inspection that the correct connector type has been employed, an analysis of the worst-case voltage drops on the power lines (particularly when the increased resistance from wire ageing is taken into account) and so on. The evidence will be captured in a C of D for the aircraft/weapon combination.

3.13   The Safety Case and Certification

For an aircraft/weapon combination to be certified for service use, there will be a number of steps taken which will ultimately capture not only the product of the integration programme but also the results of any service evaluation trials, logistic support analysis and end user constraints.

It is likely that during the integration programme, a number of weapons would have been released from the aircraft. Such releases would have been used to gather evidence to support safe separation and dispersion from the launch aircraft, to prove the ability of the aircraft to prime the weapon and to identify that the quality of the data is sufficient to ensure that the contracted integrated system engagement accuracy has been achieved. However, these trials are unlikely to have demonstrated the integrated capability in operational scenarios against real targets.

The integration contract is therefore likely to include some form of joint service evaluation, whereby the System User can use the aircraft and weapon combination in a realistic environment against targets on a weapons range. Such trials enable the System User to determine the tactics which will be used to employ the new capability in service and to identify any additional limitations which need to be placed on the use of the system when passed into the hands of squadron aircrew.

Whilst the weapon may already be in operational service on another aircraft, its integration with a new aircraft will need to be assessed for continued safety when exposed to the new environment. For example, the environment that would be anticipated on a supersonic platform will be very different from that of a subsonic aircraft, and this will have implications for the rate in which the energetic components of the weapon degrade and, therefore, the safe life of the weapon. Different nations have different criterion for determining munition safety during the whole manufacture to disposal sequence. Within NATO nations, Allied Operating Procedure (AOP) 15 (Guidance on the Assessment of Safety & Suitability for Service of Non-Nuclear Munitions for NATO Armed Forces) is the prevailing document, whilst in the United Kingdom, Joint Service Publication (JSP) 520 is the governing management system. These processes require a munition to be tested and assessed for suitability when exposed to the representative environment and, in doing so, to identify the limitations of use of the weapon when employed on the aircraft with which it has been integrated.

The total integration activities will therefore produce large amounts of evidence which will be used to support an argument that the new combination of aircraft and weapon is safe to operate within the bounds defined in the formal Release to Service (RTS). A key part of such an argument is the Safety Case.

The Safety Case is constructed as a reasoned argument with supporting evidence generated by following a recognised system safety process (capturing the results of the various safety analyses undertaken such as FMECAs, FTAs, etc.). For every hazard that has been identified at the outset of the programme, there will be supporting evidence and a narrative explaining why the hazard has been adequately mitigated. Whether this is by features in the system design, special procedures that have been instigated to control safety or by operator training, the Safety Case must capture all the arguments and evidence such that independent scrutiny of the data and associated narrative can unequivocally agree that the system is acceptably safe within the bounds of what is reasonably practicable.

Although the Safety Case is a product that is used in the overall certification process, it should not be left to the end of the programme before being compiled. As discussed in Chapter 2, safety must be a consideration throughout the integration programme. Therefore, the Safety Case should evolve during the programme, drawing on any previous Safety Cases which may exist from previous related programmes and progressively develop the structured argument that the system is adequately safe.

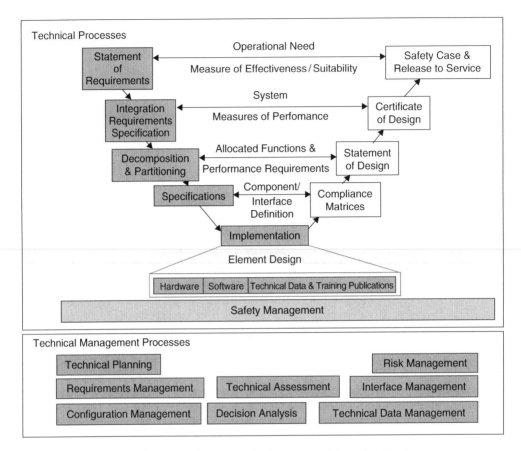

Figure 3.5   Final V-Diagram for the weapons integration process

The Safety Case will contain all the necessary information required to assess the safety of the system and will contain evidence from the product design and proving activities and the processes which have been adopted to deliver a safe system. Ultimate responsibility for the Safety Case will lie with the Contracting Agency even though one of the industry parties may have been contracted to produce it. For very complex systems, the overall Safety Case may capture significant inputs from lower level subsystem Safety Cases and integrate these into an overall safety argument.

It is therefore important to understand at the outset of the integration programme the arguments which will ultimately be used to substantiate that the delivered capability is safe. It is also important to recognise that various stakeholders may have a responsibility for delivering evidence which supports the argument, and for this, the dependencies must be identified and agreed.

The on-going activity throughout the integration programme is therefore to ensure that the relevant evidence to support the argument is being produced and documented. For example, a planned test of a particular system element that contributes to overall safety has been undertaken and has actually produced the evidence that supports the argument that the element has been implemented with the required level of integrity. Ultimately, the Safety Case should provide evidence that a recognised, rigorous safety management process has been employed; all applicable legislation and regulations have been complied with; and a valid and traceable set of safety requirements have been derived and then implemented and that hazard and risk management has been undertaken to reduce any residual risks to a tolerable level and that the evidence demonstrates that all safety requirements have been satisfied.

Should an overall systems engineering process be applied to the weapons integration programme, then a tolerably safe new capability should be deliverable to the end user within the budget and timescales agreed with the Contracting Agency at the outset.

The final version of the V-Diagram is shown in Figure 3.5.

Further Reading

North Atlantic Treaty Organisation. (2009) Guidance on the Assessment of Safety & Suitability for Service of Non-Nuclear Munitions for NATO Armed Forces. *Allied Operating Procedure 15*, North Atlantic Treaty Organisation, Brussels.

United Kingdom Ministry of Defence. (2005) Ordnance, Munitions and Explosives Safety Management System. *Joint Service Publication 520*, Defence Ordnance Safety Group, Bristol.

United States Department of Defense. (1985) Military Standard, Technical Reviews and Audits for Systems, Equipments and Computer Software. *Military Standard 1521B*, US Department of Defense, Philadelphia.

4

Armament Control System and Global Positioning System Design Issues

4.1 Chapter Summary

The control of weapons by an aircraft requires several subsystems to interact such that the weapons can be safely released and then guide to their target. The overriding principle for the control of a weapon must be such that the safety of the crew and the aircraft is not compromised. When considering safety and certification, it is essential that an aircraft only releases a weapon when intended. This would appear to be an obvious requirement, but it is the primary driver in the design of the Armament Control System.

Central to achieving compliance with this requirement is the SMS. The SMS manages the weapon load-out and controls the safe arming, release, jettison and operation of any store loaded on the aircraft, including the generation of high-integrity data messages required by the weapon to ensure its safe operation. This chapter will explore the design considerations for an SMS and its associated components and outline common system architectures that are found in a modern Armament Control System.

In designing a safe and available SMS, there is the need to consider how aircrew training will be undertaken. Once a weapon has been integrated with an aircraft, it can be very expensive to undertake aircrew training solely by firing real weapons against targets on a weapons range. To overcome this expense, various forms of training systems have been developed. This chapter will review the training aids (specifically for Air-to-Ground weapons; training for Air-to-Air weapons will be discussed in Chapter 5) that are designed into the aircraft system and discuss the system implications of each.

Although safe control of weapons is essential, the weapons also have to be released such that they have the best opportunity of hitting the target. Therefore, navigation accuracy is an important factor. The emphasis placed on the employment of GPS in guided weapon systems has enabled the continued potency of legacy aircraft and defined the baseline capability for

Aircraft Systems Integration of Air-Launched Weapons, First Edition. Keith A. Rigby.
© 2013 John Wiley & Sons, Ltd. Published 2013 by John Wiley & Sons, Ltd.

new platforms. The integration of such weapons brings with it special problems for the aircraft systems integrator. This chapter will also outline the basic operation of the GPS and discuss a number of aircraft system design issues that need to be considered when integrating such weapons.

4.2   Stores Management System Design

The safe arming, release, jettison and operation of stores are paramount in a weapons integration programme. This means that special attention must be given to the parts of the aircraft system which are involved in these functions. Within an Armament Control System, this is primarily the SMS.

The SMS is responsible for compiling and managing the stores inventory. Many smart weapons have the ability to tell the aircraft what type of weapon they are, potentially simplifying the logistics of preparing an aircraft for a mission by enabling the aircraft to automatically identify exactly what is loaded on each station. In addition to keeping the crew informed of the status and availability of weapons, this data may be required by the aircraft's Flight Control System to alter performance parameters. Also as stores are released, it may be important for continued controlled flight that the distribution of heavy stores is controlled such that their release does not impose an unstable condition on the aircraft (e.g. many heavy stores present on one wing with very few loaded on the other).

By knowing which stores are loaded on which stations, the SMS is able to ensure that only stations carrying weapons are included in any weapon release sequences whilst providing facilities for the safe (unarmed) jettison of stores if required, for example, under emergency conditions (e.g. an engine flameout on take-off, when the mass of the aircraft may need to be quickly reduced). The SMS will also control the priming of weapons prior to release in addition to ensuring that during a release sequence, safe intervals are maintained between individual releases.

All these demands make a modern SMS a complex subsystem. Multi-channel systems that are designed to maintain integrity and availability are common. Such systems employ high-integrity software and are generally designed to be immune to electromagnetic interference to ensure the system remains safe at all times when in its operational environment. However, the total system involved in the management and deployment of weapons will also consist of a number of other components such as dedicated power supplies, cockpit switches, wiring and store carriage and release systems. Together, all these components make up the platform's Armament Control System.

4.2.1   SMS Design Requirements

Whilst there are a number of weapons integration requirements partitioned to the SMS, the primary consideration is safety. Whilst it is relatively easy to design a fail-safe system, ask any pilot what would be their biggest concern having battled through a range of air defences to get to the target and it would be the inability to release the weapons. Therefore, a real SMS will have the added complexity needed to ensure that the system is available to use when required.

These two primary requirements (safety and availability) provide an apparent contradiction which the SMS designer must overcome in the system design.

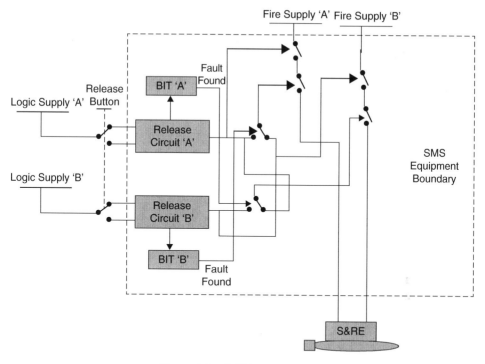

Figure 4.1   SMS implementation

In the United Kingdom, these basic requirements are captured by DEF STAN 00-970 (Design and Airworthiness Requirements for Service Aircraft) as follows:

1. The armament system shall be such that no single fault or failure shall adversely affect the safety or operation of the system.
2. The armament system shall be such that a single fault or failure shall neither:
   (i) Prevent release or jettison of the store(s) when required.
   (ii) Result in inadvertent or uncommanded release or jettison of the store(s).
   (iii) Prevent the weapon being released live and in the correct condition when required.
   (iv) Result in the arming of a weapon before release.
   (v) Prevent the weapon being made safe after having been selected live.

In order to successfully implement these requirements, some elements of the system will have to not only be duplicated, but also be cross-coupled to account for system failures. Figure 4.1 depicts a system implementation which satisfies the safety and availability requirements.

The figure shows a simple weapon loaded to a carriage system (see Section 4.2.2.4 for an overview of Suspension and Release Equipment – S&RE) which has two inputs, either of which can initiate the release mechanism. Therefore, to control the S&RE, this system implementation has two channels: A and B. The system is powered by two separate power supplies (also A and B), and these are further subdivided into Logic Supplies (used to power the SMS electronics) and Fire Supplies (used by the SMS to initiate the release of the weapon). The SMS release circuits are also duplicated and have separate inputs from a double pole changeover switch.

Tracing the circuit through, it can be seen that each release circuit controls its own channel's upper Fire Supply switch (this could be a relay which provides an air-gap in the firing chain for greater integrity or a semiconductor switch) and the other channel's lower Fire Supply switch. Therefore, if the Release Button is operated, then under failure-free conditions, both channels will energise the Fire Supply switches and the S&RE will be initiated by both channels.

Also shown in the diagram are the Built-in Test (BIT) circuits A and B. These will be monitoring the Release Circuits for correct operation, and should a failure be detected in say channel A, then the BIT A circuit will detect this and switch over control of the channel A upper Fire Supply switch to channel B. Channel B now has full authority to fire a single channel of the S&RE to release the weapon.

Of course, this implementation is simplified to demonstrate the principle and does not, for example, show how a single mechanical failure of a pole of the Release Button would be covered. In a real SMS, the BIT circuits would also monitor all critical switch inputs and outputs and use techniques such as voting to determine if a failure had occurred. Armament systems and particularly the SMS are normally designed to be tolerant to a single failure. However, this does mean that once a single failure has occurred, the system could be less safe (albeit, still available). In practice, system integrity is usually maintained by defining operating procedures that must be followed should a single failure occur.

The SMS must also display status information such as inventory, weapon status and so on and enable selections to be made such as the weapon type to be released, its release parameters such as the setting of release intervals, the selection of stations for jettison and so on. On older systems, the SMS may have associated dedicated control and display units for interfacing with the aircrew. On modern fast jet aircraft with glass cockpits, the displays and controls are integrated into the main human machine interface with the only dedicated armament components being the external switches mounted in the cockpit (see Section 4.2.2.3).

4.2.2   Other System Components

A real Armament Control System will require not only the SMS but a number of other components which, when connected together, provide a complete and safe system. The key items are:

 (i) Armament power supplies
 (ii) A Master Armament Safety Switch (MASS)
 (iii) Cockpit controls (e.g. Late Arm, Weapon Release/Trigger and Jettison switches)
 (iv) S&RE.

Each of these components will be described in the following sections.

4.2.2.1   Armament Power Supplies

Modern armament systems are electrically powered. In the United Kingdom, DEF STAN 00-970 contains a number of requirements that are identified to help support the fundamental requirements of safety and availability. For example, under emergency operating conditions, there may be a need to jettison stores. This drives both a platform safety requirement (to be able to release stores in an emergency) and the availability requirement that even under a single failure condition, the system will still operate. Dual power supplies are therefore required, and as noted earlier, these are usually subdivided into Logic and Fire Supplies. It is normal practice to provide the channel A and B supplies from separate bus bars and to route these through the aircraft whilst maintaining segregation wherever possible. This will provide a measure of fault-tolerance particularly if the aircraft sustains battle damage.

Tolerance to faults also means that it is not uncommon to adopt other design principles to reduce the potential impact on the supplies from other aircraft system failures. Therefore, the armament power supplies should be as far as practicable isolated from the power supplies to other aircraft systems. This will then give some immunity to failures that affect the power system. However, it is accepted that there may only be a single generator for the power system, so this mitigation can only be taken so far.

Whilst it is acknowledged that the Armament Control System is a critical system, it is not the only system on the aircraft which must be tolerant to power supply failures. It is for this reason that one aircraft bus bar usually has a battery that can be connected as a back-up supply in the event of an emergency (e.g. in the event of a total power generation system failure). Whilst a battery is a finite power source, under such emergency conditions, it is likely that stores would be jettisoned immediately.

UK requirements also dictate that armament wiring should not normally be formed into cable assemblies with wires that are associated with other systems. This is to ensure that damage to cable assemblies in non-critical systems cannot affect the operation of the Armament Control System. This design measure means that the routing of cable assemblies in the airframe must be carefully planned, but even so, there will be challenging installation problems to be solved particularly in confined spaces such as in the aircraft wings where the opportunities for segregation of cable conduits is diminished.

4.2.2.2 Master Armament Safety Switch

The MASS is a power isolation switch located in the aircraft cockpit. Provision of the switch is primarily a requirement to ensure the safety of ground personnel when an aircraft is armed. Originally, the switch was used only to isolate the Fire Supplies, but current UK requirements dictate that a three-position switch is used so that the SMS Logic Supplies can also be isolated. The three positions of the switch in order of rotation are Safe (when all system power supplies are isolated), Standby (when system Logic Supplies are switched on but Fire Supplies remain isolated) and Live (when both Logic and Fire Supplies are energised).

Power supplies to armament systems are required to be tolerant of high transient loads (e.g. providing the currents required to fire S&RE pyrotechnic cartridges when releasing a weapon). Therefore, the MASS does not usually carry the full current used by the armament system but, instead, switches the supplies to the coils of contactors (Armament Safety Break Contactors). It is not uncommon for such contactors to have a continuous current rating of greater than 100 A. The contactors are used to route the Fire Supplies to separate distribution bus bars that are reserved for the safety-involved armament supplies.

As noted earlier, the MASS is primarily a ground personnel safety aid. Therefore, it is essential that ground personnel know the state of the switch. Legacy aircraft have a mechanical flag used to indicate the position of the MASS, which is visible from a distance. Modern aircraft use indicator lights mounted external to the cockpit which can be viewed from a distance under all lighting conditions.

4.2.2.3   Cockpit Controls

Executive control of the armament system is always vested with the aircrew, and this is achieved by a number of switches and controls located in the cockpit. We have already discussed the MASS, but the other switches are Late Arm, Weapon Release/the Trigger and Emergency and Selective Jettison buttons.

The Late Arm switch (on aircraft in the United States, this is known as Master Arm, not to be confused with the function of the MASS), is usually a guarded toggle switch. Late Arm provides a key function in the integrity chain of the armament system, being the last 'enable' in the overall Fire Supply switching sequence, and would normally only be switched to the live position just prior to weapons release.

To initiate a weapon release, the aircrew will use the Weapon Release Button or the Trigger (for forward-fired munitions). Both these controls will usually be located on the pilot's control column and may also be guarded to avoid inadvertent operation. The Weapon Release Button is usually the final sanction to the system that the weapon package can be released from the aircraft. However, the timing of the release and the order in which weapons are released are controlled by the SMS, possibly with precise inputs from the Weapon Aiming Computer to ensure an accurate delivery of ordnance on target. The Weapon Release Button is, in reality, a commit button which enables the system to accurately deliver ordnance.

The Emergency (clear all stations) and Selective Jettison (clear crew-selected stations) buttons are used to initiate stores jettison sequences when required.

The crew must also have weapon selection and status information displayed, so that full end-to-end control of the system can be maintained. On legacy systems, there could be dedicated control panels which provide the aircrew with the ability to select weapons into attack packages and to set key parameters such as modes, fuze settings and release intervals. On modern aircraft with glass cockpits, then the weapon parameter selections would be achieved via one of the aircraft's Multi-function Displays. During an attack, key timing cues could also be displayed on the pilot's Head-up Display.

4.2.2.4   Suspension and Release Equipment

S&RE are used to attach the weapon to the aircraft. For Air-to-Ground weapons and other stores, there is a need to employ standard mechanical interfaces, which within NATO are a 'hook and eye' attachment defined by STANAG 3726. The eye or bale lugs are screwed into the store, and the aircraft mounted rack contains hooks to grab the bale lugs. Sway braces which tighten against the store are employed to reduce lateral movement of the store during captive flight.

Various types of S&RE are available. Some are gas operated, employing pyrotechnic cartridges that produce high pressure hot gas to open the mechanism and to charge pistons which push the store away from the aircraft at release. Others, particularly on systems where the speed of the aircraft at store release is relatively low or where the store is light-weight, contain an electromagnetic solenoid to operate the mechanism, with the store separating under

Figure 4.2   S&RE located in an internal weapons bay

the influence of gravity. Whilst the electromagnetic release unit requires low maintenance, the pyrotechnic cartridges used in a hot gas system produce corrosive residue which brings with it maintenance penalties. For this reason, cold gas systems have been developed. Such systems use compressed gas (usually purified air), either held in a local accumulator in each rack or in a central accumulator feeding all the racks on the aircraft.

A typical example of S&RE is shown in Figure 1.5.

Aircraft which carry weapons internally will have several pieces of S&RE fitted inside the weapons bay. Figure 4.2 shows a typical example of a weapons bay containing multiple S&RE.

Some Air-to-Air missiles employ a launcher system as the S&RE. Air-to-Air missiles can either be launched by ejection (from special S&RE designed for such weapons) or forward-fired along the launcher's rail. For ejected missiles, there are currently no standardised mechanical interfaces, making each launcher different and generally bespoke to the launch aircraft/missile. The mechanical attachments for rail-launched missiles are defined by STANAG 3842AA.

4.2.3   Typical System Architectures

Different aircraft have adopted different architectures for the Armament Control System. Figure 4.3 shows a typical architecture which could be found on a legacy aircraft.

In this example, the heart of the SMS is the Stores Management Processor (SMP). It is the SMP which contains all the functionality for controlling the weapons, be they ballistic bombs or smart guided weapons (note that in some systems, the SMP may control all safety-critical functions but may partition the mission-critical functions between the SMP and the Mission Computer).

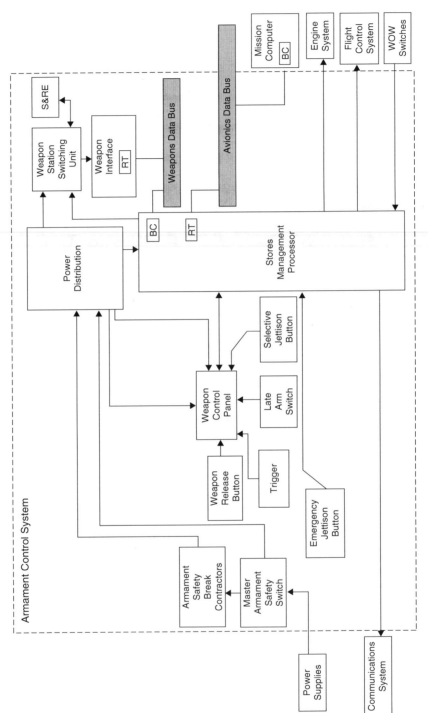

Figure 4.3 Legacy armament system implementation

Associated with the control and selection of weapon parameters such as weapon types, release intervals and fuzing options is a separate dedicated Weapon Control Panel (WCP). This also provides an interface into the SMP for the cockpit-mounted switches, with the exception of the Emergency Jettison button which connects directly into the SMP so that stores can be cleared from the aircraft even if the WCP has failed.

The SMP also interfaces with a separate Weapon Station Switching Unit (WSSU) which contains all the power switching for release and fuzing supplies to the S&RE. The WSSU could be a simple relay box or it could be a more complex safety-critical switch with multiple outputs controlling a number of separate S&RE.

In this example, the SMP is connected as a Remote Terminal (RT) to the Avionics Data Bus which is controlled by a Bus Controller (BC) located in the Mission Computer. This interface enables the SMP to receive data that will be required by the weapons such as aircraft velocities, target data and GPS data. The SMP would then format the data into weapon-specific messages which are then transmitted to the weapons over a Weapons Data Bus (with the SMP acting as the BC and the weapons providing RTs).

The SMP also has interfaces with the Weight-Off-Wheels Switches (WOWS) so that the system can determine the flight status of the aircraft (required as a safety interlock in the release of weapons). Other typical interfaces will be with the Power Distribution system, the Communication System (for providing audio signals generated by some short-range Air-to-Air missiles to the crew), the Engine System and the Flight Control System.

Whilst Figure 4.3 is typical of a legacy system, the current generation of fast jet aircraft has adopted Multi-function Displays and control panels to reduce crew workload and to efficiently use the space provisioned in the cockpit for control panels. This change in the overall aircraft system architecture has driven the migration of some armament functions from dedicated control panels into the main mission system. Such an architecture is shown in Figure 4.4.

In this example, the SMS is now a single unit in the system (the SMP). The WCP has been replaced by an integrated 'Hands on Throttle and Stick' (HOTAS – which puts all essential switch selections in easy reach of the pilot) and the Multi-function Display (which have weapon pages for providing weapon function selection and display). Much greater use is made of the Data Bus and dual Display Mission Computers (DMC) not only providing data to the SMP but also scheduling the data bus transaction to enable the SMP to communicate with the weapon stations and the Flight Control System. The SMP also controls the S&RE directly without the need for a separate WSSU. Other interfaces such as those with the Communications Systems, Engine Control System and the MASS/Armament Safety Break Contactors remain the same as for the legacy system.

These system architectures provide two possible examples. However, in reality, the actual system architecture will be defined to suit the needs of the platform and, therefore, any 'real' system may employ aspects of both architectures. For example, the amount of data being transmitted on an Avionics Data Bus in an integrated architecture may be such that there is little capacity for the timely transfer of weapon-specific data. This would then necessitate a separate Weapons Data Bus being introduced as shown in the legacy system architecture.

4.2.4   Training System

When a new weapon is integrated with an aircraft, there is a need for aircrew to be trained in its use. Initial training will be in ground-based simulators before progressing to training in an

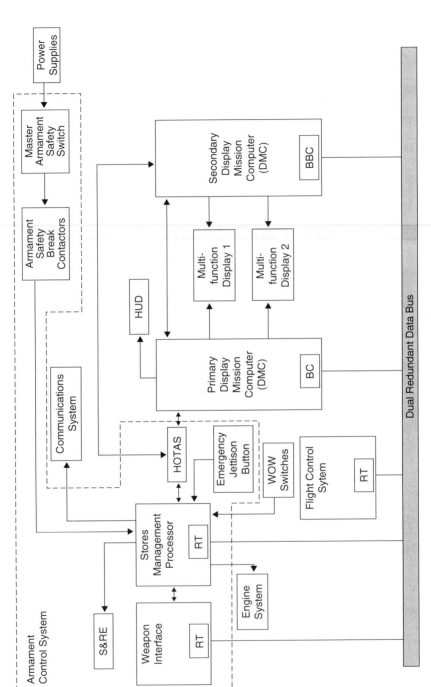

Figure 4.4 Integrated armament system implementation

aircraft. Whilst individual air forces will employ their own training methods, there will be a need for the ADO and WDO to provide an integrated capability which supports these training needs. Training must mimic the use of the operational weapons as closely as possible. Therefore, much of the implemented system functionality of the operational system should be employed with any differences being minimised to those necessary, for example, to comply with safety requirements.

For the purpose of this chapter, training for Air-to-Ground roles will be considered and the system implications for including the required capabilities such as the use of practice weapons (weapons which mimic the functionality of the operational weapon, employing the same interface with the aircraft) will be discussed.

When releasing Air-to-Ground weapons, the launch aircraft is likely to be in a relatively benign part of its flight envelope. Historically, the need was for the pilot to manoeuvre the aircraft such that when released, the bomb would follow a ballistic trajectory to the target. Interaction with other assets was minimal therefore enabling the actions leading to weapon release being well defined and mechanistic.

However, to actually release a live weapon would mean that a designated range would be required. Releasing weapons also carries a financial penalty and, with a finite weapon load-out, training duration would be dictated by the number of weapons that were carried. For training which required the release of multiple weapons, then the number of passes over the bombing range could be severely limited.

As the complexity of weapon aiming systems increased and became largely implemented in software, this provided the opportunity to build in simple training aids. Here, the system is put into a simulated weapons mode where the Weapon Aiming Computer and the SMS would enable all selections for a particular type of weapon to be made, just as the crew would for the operational weapon, with the exception that weapon release is inhibited. Being in a training mode, the SMS will enter a 'weapon released' mode after the simulated release has been completed, thereby enabling displays and system modes to be updated accordingly.

This approach means that when a new weapon is integrated with the aircraft systems, all functionality is employed for training with the exception of actually releasing a live weapon. Typically, the only real difference would be the selection of the training mode (often done by entering a special code into the system pre-flight). A significant benefit of a simulated mode (where nothing is actually released from the aircraft) is that an attack can be prosecuted against any target such as a bridge, a particular building or an airfield runway without the need to be on a designated weapons range. This provides significant flexibility for the training programme and with an inexhaustible load-out of simulated weapons means that training sorties are only limited by aircraft fuel load and crew fatigue.

Although the use of simulated weapons is an expedient and cost-effective way of introducing a training capability, there are potentially a number of pitfalls which the system designer needs to avoid.

As training is meant to be as realistic as possible, then all mode selections including weapon release need to be identical for both the live weapon and the simulated weapon. However, this could lead to the aircrew becoming confused over which type of weapon was being selected, and this could lead to a live weapon being released instead of a simulated weapon. Such a scenario should be identified during the hazard analysis at the start of the integration programme and there-fore can be mitigated. In line with the recommended hierarchy for hazard mitigation discussed in Chapter 2, mitigation should be by design such that it is not possible to access the simulated weapons whilst live weapons are present. Should this not be possible, then there should be a physical safety device such as a circuit breaker within the system that interrupts the SMS Fire Supplies when a simulated weapon is selected. However, should this be an interlock operated by the aircrew, then it should be recognised that this will introduce an opportunity for human error to occur which could result in an inadvertent weapon release. Also the use of such an interlock will mean that there is actually a different sequence of switching and moding required for the operation of a live or a simulated weapon, thereby creating a sub-optimum training facility.

A correctly designed simulated weapon training facility can provide a significant level of training opportunities and can easily be implemented such that many different weapon types including simulated smart Air-to-Ground weapons can be employed on a single training sortie.

Whilst the majority of training may be undertaken using simulated weapons, there may also be a need to sometimes attack a target on a range with a real weapon, thereby bringing all the training together in an end-to-end demonstration of aircrew proficiency. This could be achieved using operational weapons although this can be a very expensive option. For weapons with large warheads (i.e. they have a high kinetic energy), this would also mean using a range with a large safety trace area to ensure that any inaccurate releases or the release of a weapon which suffers a failure post release does not cause a hazard to ground installations or personnel.

A cost-effective solution is to use practice weapons which are generally light-weight and have similar ballistic characteristics but may not include an explosive charge (although some

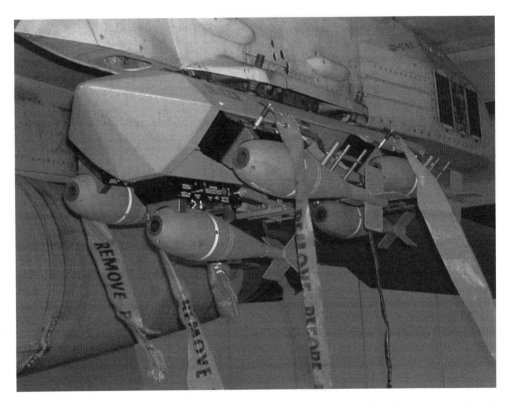

Figure 4.5 Four-station practice bomb carrier fitted with practice bombs. Reproduced with the permission of EDO MBM Technology Limited

practice bombs do include smoke generators to aid target spotters in identifying where the bomb impacts). Advantages of employing smaller practice weapons is that drag may be significantly reduced, thereby improving sortie duration, and more weapons can be carried, particularly if multiple practice weapons are fitted to a special carriage system.

A typical example of practice bombs fitted to a four-station carrier is shown in Figure 4.5.

For end-to-end training of the release of heavy weapons or weapons that have significant drag, the use of practice weapons may not provide aircraft handling of sufficient fidelity. For example, the release of a 1000 kg class weapon could cause the aircraft flight dynamics to be altered causing the aircraft to rise. The release of a 14 kg practice weapon will clearly not cause the same effect. It is for this reason and the need for pilots to experience aircraft handling qualities with a representative full load-out that training may occasionally include the release of real weapons. Whilst this could be justifiable for simple weapons such as ballistic bombs, for more complex weapons, cost will become the dominating factor.

The Contracting Agency may therefore have a need for the WDO to provide modified operational weapons. Typical modifications could include replacing the warhead, propulsion system and electronics with ballast (to support realistic weapon simulation, but not necessarily having an ability to attack a target) or having fully operational weapons that provide full capability but with the warhead replaced by a telemetry pack.

4.3   GPS: Aircraft System Design Issues

4.3.1   GPS Overview

4.3.1.1   An Introduction to the GPS

When releasing bombs at medium altitude, there is often cloud covering the target. Rules of Engagement may dictate that the release of Laser-Guided Bombs (LGBs) may be prohibited in such conditions. The use of a 'fire-and-forget' GPS-guided weapon eliminates the need to see the target prior to launch. Also collateral damage considerations are paramount, particularly in peacekeeping missions. Again, a GPS-guided weapon coupled with a low-cost inertial navigation system can greatly improve accuracy and therefore reduce the potential for collateral damage.

GPS is a space-based radio-navigation system that was originally developed by the United States as a military force enhancement system. The system works by measuring the difference between the time of reception (as defined by the receiver's clock) of a ranging signal (transmitted by the satellites) and the time of transmission contained within the satellite's navigation data (as defined by the satellite's clock) multiplied by the speed of light. This is known as the pseudo range. The original GPS supported two standards of service, but these are now being enhanced with a third service specific for an enhanced military capability. These services are:

- The Standard Positioning Service (SPS) that is designed to provide a less than military accuracy positioning service for civilian uses.
- The Precise Positioning Service (PPS) that is available primarily to the United States and its allies as a more accurate system.
- The new M-Code service that delivers improved security and accuracy.

The GPS consists of three parts. These are the satellites, the GPS receiver and a ground-based Control Segment.

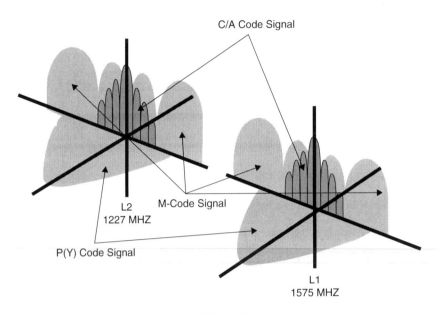

Figure 4.6   GPS signal modulation

4.3.1.2   Carrier Modulation

Each satellite transmits L-band signals known as L1 (centred at a frequency of 1575.42 MHz) and L2 (centred at a frequency of 1227.6 MHz). The ranging signals are modulated on to these carrier signals as shown in Figure 4.6, with the two carrier components (Course Acquisition (C/A) code and the Precision (P(Y)) code) in phase quadrature with each other.

Each of the original GPS carrier components are Bipolar-Phase Shift Key (BPSK) modulated by a separate bit-train. One bit-train is the Modulo-2 sum of the C/A pseudo random noise (PRN) code and the navigation data, while the other is the Modulo-2 sum of the P(Y) PRN code and the navigation data. The satellite modulates the code sequence resulting from the combining of the navigation data with the two PRN onto the L1 carrier to create a spread spectrum ranging signal, which is then broadcast to the user community.

The L2 link is used to transmit a second ranging signal that supports the original precision navigation capability for military users. The L2 carrier is again modulated by a combination of the PRNs and the navigation data to produce a spread spectrum signal. By using both L1 and L2, the receiver is able to determine the ionospheric delays and therefore improve the pseudo range estimates.

To maintain compatibility with fielded systems, the M-Code signal uses a binary offset carrier signal and spreading code so that it can co-exist with the C/A and P(Y) codes without causing interference.

The L1 and L2 signals are transmitted with enough power to ensure a minimum signal power level of −160 dBW (for L1) and −166 dBW (for L2) at the earth's surface. The different encoding scheme of the M-Code signal enables the signal to be transmitted at a higher power of −158 dBW without causing interference to the other signals and to improve the jamming resistance of the new service.

4.3.1.3   Standards of Service

SPS is provided by the L1 frequency that is transmitted by all satellites and contains the C/A code and the navigation data message. SPS is specified to achieve a horizontal accuracy of 300 m and a vertical accuracy of 500 m for 99.99% of the time at any point on the earth's surface. Better accuracies will be realised but for a reduced percentage of the time. However, from practical experience with modern receivers (e.g. hand-held satellite navigation systems), a significantly better level of accuracy appears to be the norm.

In suitably equipped receivers, the C/A code is used primarily for acquisition of the P (or Y) code (denoted as P(Y)), where the Y-code is an encrypted version of the P-code to protect against spoofing (the deliberate transmission of incorrect GPS information by a party wishing to render the system unusable). The Y-code is used in place of the P-code whenever the Control Segment activates the Anti-Spoofing mode of operation. P(Y) code receivers are designed to accept cryptographic keys in order for the United States to restrict access to the service to military systems of its own nation and those of its allies. The horizontal and vertical accuracy of navigation data produced by PPS is significantly better than that specified for SPS.

4.3.1.4   Satellite Acquisition

As the received satellite signal levels are below the level of thermal noise, the receiver uses correlation techniques to obtain the navigation signals. The receiver determines which satellites are visible from its knowledge of its present position, velocity and time (PVT) and any stored almanac data. If inadequate data is available to the receiver, it must 'search the sky' in an attempt to randomly locate and lock onto any available satellite. Providing a receiver can estimate which satellites are visible, then it can target a particular satellite to track.

The C/A and P(Y) codes generated are precisely predictable relative to the start time of the code sequence. A GPS receiver can therefore replicate the same code as the satellite. The amount that the receiver must offset its code generator to match the incoming code from the satellite is directly proportional to the range between the GPS receiver antenna and the satellite. The receiver must know GPS time very accurately as the satellite signals indicate to the receiver the time of transmission from the satellite. It is the time delay from transmission to receipt that defines the pseudo range.

The received waveform is equivalent to that of the carrier frequency spread by a regular square wave function with P(Y) and C/A code modulation. When the spread spectrum signal is received, the signal power is spread out over such a large bandwidth that the GPS signal is below the level of thermal noise. When the received signal is multiplied with the receiver-generated C/A and P(Y) codes, the satellite signal is de-spread into the original carrier frequency band. The signal power is then concentrated into a narrow band that is well above the level of thermal noise (see Figure 4.7).

To lock onto a satellite, the receiver employs carrier and code tracking loops. Both tracking loops need to work together. The code-tracking loop correlates the incoming code (C/A or P(Y)). The carrier-tracking loop adjusts its own frequencies (centred on L1 and L2) to take account of the Doppler shift of the incoming signal due to the relative velocities of the satellite and the receiver. The receiver uses the relative velocity with respect to four satellites to determine its own velocity in Earth-Centred-Earth-Fixed co-ordinates. The code-tracking loop is used to make the pseudo range measurements between the satellite and the receiver. The

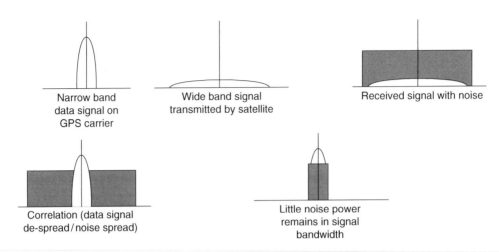

Figure 4.7 GPS signal spreading/de-spreading

code-tracking loop generates a replica code that must be aligned with the incoming code. In order to match the replica with the received signal, the centre frequency of the replica must be adjusted to the centre frequency of the incoming signal. This is assisted by the output from the carrier-tracking loop. To fully acquire the satellite, the phase of the replica code must be matched with that of the incoming code. The phase difference is therefore directly proportional to the pseudo range between the satellite and the receiver. The replica code is adjusted in phase until an exact match is achieved. Once both the carrier and code tracking loops have locked on to the received signal and the code stripped from the carrier, the navigation message can be demodulated.

During periods where the receiver may not be able to maintain code and carrier tracking (e.g. during the ejection sequence of a weapon from a launch aircraft), the receiver will normally maintain code tracking. If only code tracking is available, the receiver will slew the locally generated carrier and code signals based on predicted rather than measured Doppler shifts. These predictions can be assisted with data from other sources such as an IMU.

In order for the receiver to navigate, it has to acquire and track satellite signals to make pseudo range (distance) and delta range (integrated Doppler shift) measurements and collect the navigation data message. When a single satellite is being tracked, the receiver can demodulate the navigation message and read the almanac data for all the other satellites in the constellation. From this, it can then acquire more satellites.

Figure 4.8 provides an overview of the data contents and structure within the navigation message. The message contains 25 data pages, each page consisting of 1500 bits. Each page is divided into five sub-frames of 300 bits each. As navigation data is clocked at 50 bps, it takes 30 s to receive one data page and 12.5 min to receive all 25 data pages. Each sub-frame consists of ten words, each of 30 bits long. Sub-frames 1, 2 and 3 have identical data content on all 25 pages. This allows the receiver to obtain critical navigation data within 30 s of starting to decode data. Sub-frames 4 and 5 are each transmitted several times, providing data for the full satellite constellation.

The sub-frame data includes information to determine satellite time of transmission, satellite position, satellite health, satellite clock correction, propagation delay effects, time transfer to Universal Time Co-ordinated (UTC) and constellation status.

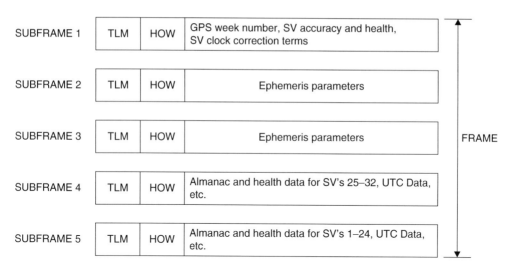

Figure 4.8 GPS navigation message content

Each sub-frame starts with a Telemetry word (TLM) and a Hand-over word (HOW) pair. The HOW is used to transition from C/A code to P(Y) code tracking.

The M-Code signal contains a more flexible data message that introduces greater flexibility and is structured to enable future changes and new features to be easily introduced.

4.3.1.5   C/A Code Acquisition

In a P(Y) code capable receiver, C/A code will normally be used to assist in the acquisition of the P(Y) code. With current almanac and ephemeris available to the receiver and a course estimate of PVT, then a Time to First Fix (TTFF) of less than 30 s is possible. Guidelines for the Global Positioning System Receiver Application Module (GRAM), GPS-GRAM-001A dated 25 February 1998, identifies a carrier-to-noise measure for the level of susceptibility to jamming of the received signals. The measure is the ratio of the de-spread signal within the receiver and the noise density in the appropriate bandwidth (2.046 MHz for C/A code and 20.46 MHz for P(Y) code) at the output of the receiver's code correlation circuits (C/N$_0$ – measured in dBHz). C/A code acquisition has the highest C/N$_0$ requirements which equates to an achievable jammer-to-signal ratio (J/S) of around 24 dB.

4.3.1.6   C/A–P(Y) Hand-Over

The C/A code repeats every 1 ms allowing for a minimal receiver search window. The P(Y) code repeats every 267 days, and each satellite is assigned a unique one-week segment of this code. Acquisition requires the receiver to have some prior knowledge about the current position in the seven-day cycle of the P(Y) code. The HOW contained in the navigation message provides the P(Y) code phase information. The receiver uses the HOW and the C/A code-derived navigation solution to minimise the P(Y) code search window. With current almanac and ephemeris available and a course estimate of PVT, then a TTFF of less than 20 s is possible particularly if a semi-accurate (accurate to within 10 ms of GPS time) time initialisation is available. The C/N_0 and J/S ratios are identical to C/A acquisition.

4.3.1.7   Direct P(Y) Code Acquisition

A suitably equipped receiver (i.e. one that is equipped to accept daily crypto-keys) can attempt to acquire the P(Y) code directly without first acquiring the C/A code. This requires a good knowledge of the receiver position and a very good knowledge of GPS time. For direct P(Y) code acquisition, the use of the Time Mark Pulse (TMP) or One Pulse per Second (1 PPS) discrete electrical input to the receiver is essential. This acquisition mode gives good protection against interference (jamming or spoofing). With precise position and velocity available to the receiver, a TTFF of less than 8 s is possible. This mode has lower C/N_0 requirements, and satellite acquisition is possible with a J/S of 34 dB. If very accurate PVT is provided to the receiver, then satellite acquisition with a J/S of 53 dB can be realised.

4.3.1.8   Direct M-Code Acquisition

Direct acquisition of the M-Code signal is possible and, with the binary offset code spreading, enables faster acquisition times (TTFF) than direct P(Y) code acquisition. The enhanced precision of the M-code data word enables greater navigational accuracy to be achieved by suitably equipped receivers.

4.3.2   Satellite Acquisition Concepts

In assessing the ability of a GPS receiver to acquire satellites and to produce a navigation solution, there are three key performance parameters which need to be considered. These are Reaction Time (REAC), Time to First Fix 1 and Time to First Fix 2. REAC is defined as the elapsed time from receiver turn-on to the display or output of present position, velocity and time with specified accuracies, derived from satellite signals using P(Y) code and current ephemeris data, assuming that current almanac data is available.

TFF1 is defined as the elapsed time from transition into track mode for the receiver to acquire the satellite signals using a C/A to P(Y) code hand-over, collect the navigation data and compute a navigation solution of present PVT with specified accuracies, under the conditions that the receiver has been on for longer than a specified period of time.

TFF2 is defined as the elapsed time from transition into track mode for the receiver to acquire the satellite signals using direct P(Y) code acquisition, collect the navigation data and compute a navigation solution of present PVT with specified accuracies, under the conditions that the receiver has been on for longer than a specified period of time.

To achieve the shortest post-launch acquisition time, the launch aircraft should endeavour to provide the GPS receiver in the weapon with the conditions necessary for TTFF2. This requires the aircraft to provide the receiver with accurate PVT initialisation information and current satellite ephemeris and clock offset data. As most aircraft have a GPS as a part of their navigation system, the weapon initialisation data downloading requirements for TTFF2 conditions are not difficult to satisfy. Besides the data requirements, either the TMP or 1 PPS signals must also be provided by the launch aircraft. However, the provision of a GPS radio frequency (RF) feed to the weapon can alleviate some of these requirements.

4.3.3   Acquisition Strategies

For weapons with a very long time of flight, it may be possible to power up the GPS receiver at the point of release and then acquire satellites without any pre-launch aiding. However, there will be a finite time for acquisition to be achieved (assuming the J/S figure is sufficiently high) before the weapon can begin to navigate. In reality, it is unlikely that a weapon would rely purely on GPS as its guidance system and would probably have a low-cost inertial system integrated with the GPS. Therefore, an acquisition strategy that employs pre-launch aiding is more likely to be successful.

For a weapon that does have an integrated IMU/GPS, then the inertial system will be powered pre-launch and some level of alignment achieved. Post-launch, the weapon will employ the inertial system for guidance in order to allow the GPS receiver to acquire the satellites. If the GPS receiver also has knowledge of satellite ephemerides (e.g. a basic set of GPS initialisation data is loaded into non-volatile memory prior to the launch aircraft becoming airborne), then an acceptable implementation of GPS can be achieved.

If a weapon's GPS receiver can be powered prior to launch, then there is the possibility of providing a direct feed of the GPS RF signal to the receiver. This could be from the weapon's own antenna or from the launch aircraft, routed through the electrical interface connector. This concept enables the GPS receiver to acquire the satellites in its own time, on route to the target area. In this case, the GPS receiver may not be subjected to jamming and will have sufficient time for its internal clock to stabilise its temperature. On release, the GPS RF signal may be lost for a short period (e.g. when the electrical interface connector breaks and prior to the weapon's antenna becoming un-obscured from the shadow of the aircraft). Having been previously tracking satellites, the time required for the GPS receiver to re-acquire the satellites will be greatly reduced.

The optimum solution is to align both the GPS receiver and the inertial system prior to launch. Here, the time on route can be used to ensure that both parts of the navigation system are aligned and are tracking each other. Post release, the GPS receiver will quickly re-acquire the satellites with the assistance of the IMU. The initial time required for the GPS to achieve its first satellite acquisition prior to launch can be reduced significantly by the use of either TMP or 1 PPS.

4.3.4   GPS Signal Distribution

4.3.4.1   Receiver Requirements

A weapon's GPS receiver must be capable of operating when it is receiving RF signals from its antenna but also (if it expects to be aligned before launch) through the aircraft electrical interface connector. Standard receivers such as those that comply with the GRAM guidelines accept that between the antenna and the input to the receiver, there will be additional signal losses incurred. GRAM assumes this loss to be of the order of 4 dB. For a receiver mounted in a weapon and receiving RF from the weapon's antenna, then this figure is realistic. However, if the weapon is located on a weapon station underneath the aircraft, or even inside a weapons bay, then the losses associated with the aircraft cables will be significantly greater than 4 dB. The aircraft cabling will also introduce additional noise.

The Signal-to-Noise Ratio (SNR) or more realistically, the Noise-to-Signal Ratio, is already such that the GPS signal level for C/A code is around 20 dB below the thermal noise floor. Additional noise introduced by the aircraft cabling will increase the REAC time of the receiver.

Correlation Loss (defined as the difference between the satellite power received in a 20.46 MHz bandwidth and the signal power recovered in an ideal correlation receiver of the same bandwidth) must also be taken into account. On the L1 and L2 channels, the worst-case correlation loss occurs when the carrier is modulated by the sum of the P(Y) code and the navigation data stream. For this case, the correlation loss apportionment is:

- Satellite modulation imperfections 0.6 dB.
- Ideal user equipment receiver waveform distortion 0.4 dB (due to the 20.46 MHz filter characteristics).

However, a total figure of around 0.5 dB can be assumed in practice.

4.3.4.2   Typical Cable Losses

There are many different cable types that could be used for routing GPS RF signals within an aircraft. MIL-STD-1760 makes provision for the High Bandwidth 1 Type B signal for routing GPS RF to stores. The standard defines the use of MIL-C-17/113 cable for this signal. Where aircraft routing networks have long cable runs, then the installed performance of the cable must be considered. Low loss versions of cable may be used, and in parts of the distribution network that do not have to mate with the small co-axial contacts in the MIL-STD-1760 interface, a higher performance cable should be used wherever possible.

The typical lengths of cable runs from the aircraft antenna to the weapon interface for three different actual aircraft are:

- Aircraft A – 12 m (giving approximately 12 dB of cable losses).
- Aircraft B – 24 m (giving approximately 17 dB of cable losses).
- Aircraft C – 15 m (giving approximately 12 dB of cable losses).

In this example, it is likely that Aircraft C employs cables of a lower attenuation than Aircraft A (the losses quoted take into account the different cable types employed by the different aircraft and are for the cable losses only).

4.3.4.3   Problems

For a passive distribution network, the problems are:

- Too little signal is delivered to the weapon's GPS receiver (for L1 C/A code, the signal level could be as low as −177 dBW).

- Poor noise performance (the SNR could therefore be around −37 dB for a thermal noise floor of −140 dBW).

It is clear from these examples that the aircraft distribution network will have to include some level of signal amplification. The obvious aim of this will be to boost the signal level at the input to the weapon's GPS receiver. However, the additional amplification will also increase the noise levels in the system. Therefore, careful design is required in order to control potential noise problems.

If amplification stages are introduced, then the designer must be aware that a standard GPS receiver may provide a dc pre-amp/antenna power supply on the RF line. This voltage could be as high as +12 V dc.

4.3.5 Aircraft Requirements

4.3.5.1 Appendix A to MIL-STD-1760

Appendix A to MIL-STD-1760 defines the requirements for routing GPS RF signals to the weapon. The requirements address several problems, namely:

- Aircraft distribution network gain.
- The noise performance of the aircraft distribution network defined as Figure of Merit (in a GPS signal distribution system, Figure of Merit is defined as $G_e - 10\log_{10}T$, where G_e is the effective gain of the aircraft distribution network in decibels and T is the output noise temperature in kelvins (Figure of Merit in a GPS distribution network should not be confused with the integer value of position errors as defined in documents such as the GRAM guidelines)).
- The signal power levels that the store can expect at its electrical interface connector.
- The SNR of the signal at the stores electrical interface connector.
- The requirements also limit the dc bias that the store can apply to the RF signal line.

4.3.5.2 Aircraft RF Routing Requirements

To satisfy the requirements of MIL-STD-1760 Appendix A, the aircraft will have to provide sufficient gain to account for the cable losses and to provide a positive boost to the signal, such that the level delivered to the store's electrical interface connector is above the minimum defined in the GPS specification. The SNR must also be controlled.

As the GPS signal levels are below the thermal noise floor, it is impossible to detect them with laboratory equipment. A method of determining the acceptability of the performance of the aircraft distribution network is to monitor the C/N_0 parameter generated by the receiver (this parameter is generally available as a data word output from receivers that employ a MIL-STD-1553B multiplex data bus interface. As a 'rule-of-thumb', a C/N_0 of 28–50 dBHz is acceptable for C/A code acquisition.

GPS signal distribution will be discussed further in Section 6.3.4.

4.3.5.3   Aiding Data

The acquisition performance of the weapon's GPS receiver can be greatly improved if aiding data can be provided by the aircraft. This will typically be provided over the aircraft's MIL-STD-1553B multiplex data bus and could consist of:

- Accurate PVT from the aircraft's navigation system.
- Lever arm data to define to the store, its displacement from the aircraft's 'centre of navigation' datum.
- Satellite data such as almanac and ephemeris.

If such data is available to the weapon's receiver, then fast acquisition times are easily possible.

4.3.6   Aircraft Implementation Concepts

The aircraft system design considerations must take account of the way different stores may use GPS. There are many weapons in service that employ a GPS receiver as a part of their guidance control mechanisms. The requirement on the aircraft system to support such weapons differs for each weapon. The aircraft designer must consider which weapons are likely to be integrated with the aircraft and determine the optimum solution that will satisfy the weapon interface requirements in a cost-effective manner. Concepts that could be adopted will now be discussed.

4.3.6.1   Satellite Acquisition Post-launch (No Pre-launch Aiding)

From an aircraft designer's viewpoint, this is the simplest implementation. It relies entirely on the weapon's GPS receiver acquiring satellites post-launch with no aiding being provided from the aircraft. The weapon may have had satellite almanac data pre-loaded into non-volatile memory pre-flight to assist in the acquisition process.

   The weapon would have to perform a C/A code acquisition and then hand-over to P(Y) code. The receiver would therefore be susceptible to jamming/spoofing, and the REAC time could be extended. A weapon that could operate with this level of interface with the aircraft would need a long time of flight in order to acquire satellites and begin navigation.

   As the majority of the current GPS-assisted guided weapons tend to have a relatively short time of flight, then this aircraft design concept is unlikely to be acceptable.

4.3.6.2   Satellite Acquisition Post-launch (with Pre-launch Aiding)

For weapons that have an integrated GPS and IMU, this concept is more realistic. The aircraft can supply power to the weapon to align the IMU and to perform pre-heating of the GPS receiver's internal clock. If a data interface is available, then satellite data (almanac, ephemeris and GPS time) can be provided to the receiver.

During the weapon release phase, the IMU will provide a measure of short-term stability and will continue to provide position and velocity data to the receiver. Once clear of the airframe, the receiver will begin satellite acquisition. In this case, a relatively short REAC time is possible (of the order of seconds assuming that the receiver has a very good estimate of GPS time). However, as the receiver is not locked to the GPS signals, then in a jamming environment, REAC could be extended (assuming that the weapon does not employ Anti-Jam techniques).

4.3.6.3 Satellite Acquisition Using RF-Only Pre-launch

For less complex aircraft, the provision of an intelligent data interface may not be possible. In such a case, a weapon that can accept GPS RF only, pre-launch (plus a power supply) could be integrated. With this concept, the weapon's GPS receiver could be powered up a long time before the release point is reached. This enables the receiver's internal clock to stabilise its temperature and for the carrier and code tracking loops to acquire satellites. As the receiver can be operating on route, then it is less likely that it will be subjected to jamming/spoofing. If the weapon has an integrated GPS receiver and IMU, then the GPS receiver can be used to align the inertial elements prior to launch. Post-launch, the IMU can aid the GPS receiver in re-acquiring the satellites. Time of flight of the weapon is not critical to this concept as re-acquisition time of the satellites' post-launch should be achieved quickly.

4.3.6.4 Satellite Acquisition Using RF Plus Aiding Data Pre-launch

Although it is a more complex concept from the aircraft designer's viewpoint, using RF and aiding data pre-launch does provide for all the common interfacing options. The concept retains all the benefits of the previous concept but enables a faster acquisition time on the aircraft. It should be noted that the performance of a multi-sensor aircraft system should enable very accurate PVT to be provided to the weapon's GPS receiver, and this will improve the satellite re-acquisition time post-launch.

4.3.6.5 The Use of Critical Timing Pulses

As mentioned previously, the acquisition time of a GPS receiver and the performance of the receiver in a jamming/spoofing environment can be greatly improved if the GPS receiver can be supplied with critical timing information. Critical timing information is essential if direct P(Y) code or M-Code acquisition is to be achieved. The GPS provisions two timing signals. These are TMP and 1 PPS. These receiver precise time interfaces are electrically different but either can be used to accurately mark time. The precise time input to the receiver represents the exact instant of change of the UTC one-second-rollover. Data over the data bus (Precise Time and Time Interval (PTTI) message) are used to identify what time it was at the UTC one-second-rollover.

As the precise time interface is used to reduce the uncertainty of the receiver's initial time estimate, TTFF is significantly reduced.

Such aiding can only be used if the PTTI message can be transmitted over an intelligent data interface to the weapon.

4.3.7   Cost of Complexity

The four concepts outlined earlier can all provide a capability to employ GPS-assisted guided weapons on aircraft. However, there is always a performance versus cost trade-off to be made.

The simplest interface (satellite acquisition post-launch (no pre-launch aiding)) could have major implications for the design of the weapon. Although it will significantly reduce aircraft integration costs, the level of performance achieved by the overall weapon system may not be acceptable.

A practical compromise is provided by the intermediate concepts (satellite acquisition post-launch (with pre-launch aiding) and satellite acquisition using RF-only pre-launch). The first avoids any problems with the routing of sub-noise level signals around the aircraft but requires added system software complexity to provide the necessary GPS aiding data. However, most of the initialisation information downloading requirements should not be difficult for typical modern aircraft to satisfy. For 'simpler' aircraft, the provision of 'RF-only' pre-launch could be a cost-effective concept.

For aircraft that already provision suitable interfaces (e.g. MIL-STD-1760) at the weapon stations, then the 'complex' interface is feasible and may be cost-effective in the long term. Such an interface will be readily adaptable for a variety of weapons that use a GPS receiver in their guidance control systems. It is a generally accepted point that once an aircraft has the capability to support stores that comply with MIL-STD-1760, then the cost of integrating future compliant stores is significantly reduced. This interface concept should not be discounted due to its complexity. It should be noted that front line aircraft currently in service in the United States and Europe already support MIL-STD-1760 interfaces and therefore the flexibility that this provides can be readily exploited. However, the exact concept to be adopted will depend on the Contracting Agency's requirements and the weapon integration philosophies of the individual aircraft project.

Further Reading

NATO Standardisation Agency. (1996) Bail (Portal) Lugs for the Suspension of Aircraft Stores. *STANAG 3726*, NATO Standardisation Agency, Brussels.

NATO Standardisation Agency. (2000) Rail Launched Missile/Launcher Mechanical Interface. *STANAG 3842AA*, NATO Standardisation Agency, Brussels.

United Kingdom Ministry of Defence. (2012) Design and Airworthiness Requirements for Service Aircraft. *Defence Standard 00-970*, Defence Equipment & Support UK Defence Standardisation, Glasgow.

United States Department of Defense GPS Joint program Office. (1998) Guidelines for the Global Positioning System (GPS) Receiver Application Module (GRAM). *GPS-GRAM-001A*, February 25, NAVSTAR GPS Joint Program Office, El Segundo.

United States Department of Defense. (2007) Aircraft/Store Electrical Interconnection System, Appendix A. *Military Standard 1760E*, United States Department of Defense, Philadelphia.

5

Weapon Initialisation and Targeting

5.1 Chapter Summary

Different weapons demand different methods of targeting. Targeting a weapon, be it an Air-to-Ground weapon or an Air-to-Air missile, can be very complex and place great demands on the performance of the aircraft systems. For the accurate targeting of a smart weapon, it is essential that the aircraft and weapon axis reference systems are initialised to provide a common reference, thereby removing position and velocity errors. This chapter will discuss weapon initialisation and examine the different ways in which weapons are targeted covering the accurate delivery of ballistic bombs, the flexibility of targeting for smart weapons and the sensors types, target prosecution strategies and training elements of Air-to-Air missiles.

5.2 Targeting

For a weapon to successfully prosecute its target, it must have a means of identifying target location and potentially the ability to make corrections for errors caused by the dynamics of weapon engagement. It is no surprise that different weapons employ different techniques for targeting, some employing navigation systems often supplemented with terminal guidance seekers. In this chapter, we will explore different ways in which weapons are targeted and how the aircraft system can influence the terminal accuracy.

There are broadly two methods available for targeting weapons, these being against pre-planned targets and Targets of Opportunity (TOO). With a pre-planned target, the target type and location are known prior to commencing the sortie, therefore the optimum weapon type, fuzing options, method of attack and approach and egress routes can be meticulously planned and rehearsed.

Aircraft Systems Integration of Air-Launched Weapons, First Edition. Keith A. Rigby.
© 2013 John Wiley & Sons, Ltd. Published 2013 by John Wiley & Sons, Ltd.

TOO are identified whilst the aircraft is in flight and therefore do not have a pre-defined mission plan to follow. The accurate location of the target must therefore be determined using on-board sensors or target co-ordinates obtained from a third party. The aircraft's Mission System must be provided with the co-ordinates of the target, so that a new route can be determined and the aircraft steered to the weapon release point. However, targeting using only the aircraft's own sensors will introduce an error between the target's actual location and its location as computed by the aircraft's subsystems. This is caused by the level of uncertainty over the aircraft's own position caused by sensor inaccuracies, computational processing rates and so on. This error is known as the Target Location Error (TLE) which is expressed as an error in three dimensions: latitude, longitude and height. TLE would usually be referenced to the co-ordinate system used by the aircraft's navigation system (e.g. World Geodetic System 1984 – WGS-84), which itself is an ellipsoidal approximation of the earth's surface. This approximation introduces a further element of error between the true location of the target and the location fixed by the navigation/attack system. Whilst a co-ordinate system such as WGS-84 includes equations that can be used to minimise the error between the idealised ellipsoid it defines and the real world, the navigation and weapon aiming systems need to be highly accurate if TLE is to be minimised.

5.3   Aiming of Ballistic Bombs

For the accurate delivery of ballistic bombs, the accuracy of the aircraft's navigation solution at the point of release is paramount. Where an aircraft uses only inertial systems to generate aircraft position data, then the navigation solution will drift with time. Whilst fixed corrections could be added to account for some of the drift, the aircrew procedures would also ensure that the navigation system is continually updated. This would be achieved by a method referred to as 'on top fix'. This method allows the crew to compare the aircraft-generated position with that of a fixed over-flown point on the ground (such as a bridge or road intersection) which has a known location. The difference between the known location and the aircraft solution when the location is over-flown is the navigational error in the system. Knowing the error, the aircraft's navigation solution can be corrected.

Ballistic bombs, after release from an aircraft, follow a trajectory defined by the laws of physics. Accurate delivery requires knowledge of how the bombs will fall, extrapolating back from the target to the release point. Whilst this may seem predictable, there are many factors which will affect the bomb's trajectory. These will include the aircraft speed and attitude at the point of release, the downward ejection force imparted by the aircraft S&RE, wind velocity and direction (which itself will vary from the release point throughout the ballistic trajectory) and the air density profile during weapon flight. The effects of aerodynamic drag and build tolerances in the bomb which affect the centre of gravity will also influence the weapon aiming solution. Much effort is therefore expended during the design of the weapon aiming algorithms to ensure that an approximation of the bomb's ballistic trajectory can satisfy all the variables such that accurate delivery can be achieved.

The navigation system must therefore determine the aircraft's exact position, altitude and three-dimensional velocities in order to minimise targeting errors. Modern aircraft will combine the output from multiple sensor systems to achieve the desired release point accuracy.

Techniques such as Kalman Filtering, in which a mathematical approach to linear filtering and the prediction of a varying signal in the presence of noise is employed, will also help to improve accuracy.

The Weapon Aiming Computer will continually compute the impact point of a bomb released from the aircraft's current location in the sky. The bomb is released when the computation identifies that the Continually Calculated Impact Point (CCIP) overlays the target's position. However, the system will inevitably exhibit processing delays, data transmission delays and system latencies that must also be accounted for in the aiming solution. The Weapon Aiming Computer may therefore advance the release point to account for these delays. Further inaccuracies can be introduced due to the aiming calculations taking a finite time to complete, whereby the point where the CCIP overlays the target exactly may occur part-way through a processing cycle. To overcome this, the designer of the weapon aiming algorithms will undertake statistical analysis to determine the error in the true release point and the first calculated release point solution so that this too can be factored into the overall calculation, thereby further minimising system-induced errors. In reality, the aiming calculation will be performed at a rate largely dependent on the processing power available in the Weapon Aiming Computer.

The actual impact point of a weapon will have a Gaussian distribution about the mean impact point with ideally, the mean impact point coinciding with the target's location. In practice it rarely does.

5.4 Aircraft/Weapon Alignment

Due to the inherent inaccuracies of ballistic bombs, modern Air-to-Ground weapons employ some form of navigation system. In the previous chapter, the strategies for employing GPS receivers in weapons were explored. This included the alignment of the aircraft and weapon navigation systems using PVT aiding data. However, it should be noted that the point where the aircraft measures its current position and velocities is different from that of the weapon, the two centres of navigation being offset due to their relative locations. This offset can be critical to the terminal accuracy of the weapon. For example, if a weapon employing a low-cost navigation system is aligned to the aircraft's navigation system without the positional offset being accounted for, then the weapon will believe it is starting its terminal flight from a position located at the aircraft's centre of navigation (usually the origin of the axis system), instead of its actual location, say on a wing station. Therefore, it is understandable that the weapon (when environmental factors which influence the accuracy of the weapon's trajectory are ignored) would miss its target by a distance equal to the three-dimensional displacement from the aircraft's axis origin. This is overcome by aligning the weapon's axis system to the aircraft's axis system, taking into account the offset.

Figure 5.1 shows the aircraft axis system as an orthogonal triad of axes X_a, Y_a and Z_a, centred at origin O_a. Similarly, the weapon will have its own axis system X_w, Y_w and Z_w, centred at origin O_w. As the weapon can be located on a number of store stations on a number of different aircraft, the displacement of its axis system needs to be referenced to a common point shared by both the weapon and the aircraft. This point is often defined as the reference point located midway between the attachment points of the store to the aircraft (denoted X_r, Y_r and Z_r, centred at origin O_r).

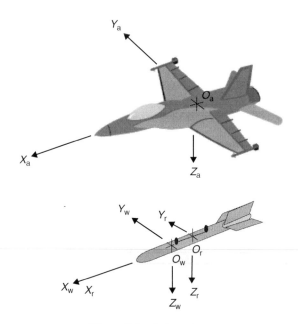

Figure 5.1   Axis systems

As the aircraft designer understands the relative offset of the reference axis centred on O_r, it is possible to transpose all position and velocity data (and any other dynamic data parameters such as inertia) required by the store as if it was being sourced at O_r, rather than at O_a.

Similarly, knowing the displacement of its own axis O_w from O_r, the weapon designer can further transpose the data presented into the weapon's axis reference system. The weapon will then accurately know its position and velocities in space.

5.5   Aiming of Smart Air-to-Ground Weapons

When released, a smart weapon, particularly if it does not have a propulsion source, will initially follow a ballistic trajectory until its guidance system takes control and begins to steer the weapon towards the target. This guidance capability of smart weapons makes it possible to trade energy at release with route to the target and desired impact conditions (e.g. to impact the target from a pre-defined direction and azimuth and elevation angles). This complicates the mission planning activity but does provide a greater level of flexibility for the attack.

Smart weapons do not have the stringent constraints on release point imposed by ballistic bombs. In effect, the aircraft need only be in a three-dimensional volume in the sky, which is a function of the release altitude, speed, distance from the target and weapon performance. This ability overcomes the need for the aircraft to calculate the CCIP with its inherent errors. The volume in the sky where, if released, the bomb will reach the target is known as the Launch Acceptability Region (LAR – see Figure 5.2).

Whilst Figure 5.2 shows the LAR in relation to a point target, it can also be shown as a projected 'footprint' on the ground (in effect, an inversion of Figure 5.2) which would show all the possible impact points if the weapon was 'launched now'.

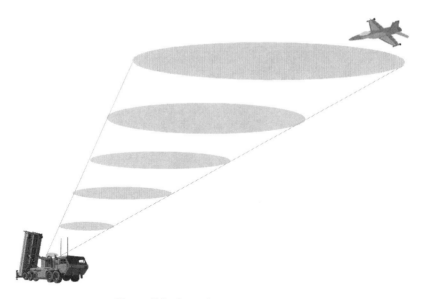

Figure 5.2 Launch acceptability region

The most accurate method of defining the LAR is to employ a six-degree-of-freedom (6-DOF) model that can predict the weapon's trajectory from release to impact for a given set of release conditions. The 6-DOF model uses the dynamics of the weapon movements (body rates, angles, etc.), the environmental conditions (temperature, air density, wind speed, etc.) and the weapon's flight control system dynamics in order to predict the flight of the weapon. The 6-DOF model is considered to provide a true representation of the dynamics of the weapon. However, the continuous processing of a 6-DOF model in an airborne computer would be prohibitively complex. It is for this reason that the models are simplified to give a good approximation of the true weapon performance. However, even a reduced fidelity LAR requires continuous calculation which can place significant demands on the Weapon Aiming Computer's processing power, particularly if a number of smart weapons are to be released in a single attack against dispersed targets. In such cases, multiple LARs would be combined with the resulting LAR being of a significantly reduced volume, but would accommodate the requirements for all weapons in the attack. The update rate of the combined LAR will become critical as minor changes in parameters for all individual LAR computations could cause dramatic changes in the combined LAR causing great variations in the displayed LAR.

Two methods are commonly used for calculating the LAR, these being the dynamic LAR and the parametric LAR. A dynamic LAR employs the equations of motion based on simplified physical characteristics of the weapon in a similar way to a 6-DOF model, but usually with a reduced number of degrees of freedom, typically a 3-DOF model. The parametric model is matched to the output of the 6-DOF model but uses an approximation such as 'least squares fit'.

In order to provide cues to the aircrew, the display of the LAR is often simplified as shown in Figure 5.3. In this example, markers for the maximum and minimum boundaries of the LAR are displayed (with the plane of the LAR also indicated to show how the markers are constructed). LAR cross-track limits may also be displayed in order to improve overall situational awareness as would the aircraft's present position in relation to the LAR.

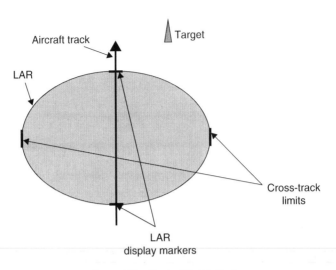

Figure 5.3 Typical LAR display

It is not uncommon for a Weapon Aiming Computer to hold several different LAR algorithms for different weapons. However, it is desirable to have a single LAR algorithm which can be used for any weapon purely by changing the coefficients of the equations used. Such an approach has been developed by the Society of Automotive Engineers (SAE – a commercial technical standards organisation) and is documented in a number of Aerospace Information Reports (AIR5682, AIR5712 and AIR5788). These are discussed in more detail in Chapter 7.

5.6 Air-to-Air Missiles

5.6.1 Sensors

Air-to-Air missiles provide a more challenging targeting situation as the target is likely to be moving rapidly and perhaps performing evasive manoeuvres. A number of engagement strategies can be employed with the actual method dictated by the type of sensor available. The most common sensor types are IR and radar although some missiles employ both types.

Early Air-to-Air missiles employed a single-element IR sensor to detect the heat emitted from an enemy aircraft's engines. This relatively simple technology would detect any source of heat including countermeasure flares or the sun, causing the missile to lock-on to a false target. Although these early seekers were improved, modern IR sensors employ IIR focal plane arrays which enable the missile sensor algorithms to discriminate between real and false targets by analysing the image captured with known target signatures.

As IR sensors do not provide significant detection range, then missiles which employ this technology can only engage targets at a relatively short range, thereby putting the launch aircraft at risk of counter engagement. Sensors employing radar technology have a significantly greater range and therefore afford the ability to engage targets at a correspondingly increased range. However, with the increased use of defensive mechanisms, radar sensors employ strategies such as 'home on jam', where the sensor may act in a passive mode to identify the relative bearing of jamming signals. The missile can then manoeuvre to attack the jamming source.

5.6.2   Engagement Modes

There are two primary engagement modes in which Air-to-Air missiles can be used, these being 'Lock-on-Before-Launch' (LOBL) and 'Lock-on-After Launch' (LOAL). In a LOBL engagement, the missile seeker is locked on to the target prior to being launched. For a short-range missile with an IR sensor, this could be achieved by simply manoeuvring the aircraft so that it is pointing directly at the target (so-called boresight). However, trying to achieve sensor lock-on whilst the target is itself manoeuvring requires the pilot to use his flying skills to keep the target within the sensor's field of view. A practical aid for the pilot is for the aircraft system to use scan patterns, where the IR sensor is moved in a defined pattern to search for a potential target. On detection of a target, missiles such as the AIM-9 Sidewinder generate an audio tone which is passed through the aircraft's Communication System to the pilot's headset. The audio tone can be used by the pilot to judge if the missile's sensor has acquired the target. When it has, the missile sensor can then be locked on to the target causing the audio tone to be modulated by the missile to inform the pilot of missile lock status.

The boresight and scan pattern methods of aiming short-range missiles can be ineffective against highly manoeuvring targets. To overcome this, missiles such as the Sidewinder employ an extended acquisition mode, whereby the IR sensor can be slaved to a target bearing generated by the aircraft's radar or Infrared Search and Track (IRST) system. This enables high off-boresight target acquisition to be achieved, constrained only by the gimbal limits of the seeker. However, even this is sometimes insufficient to engage the target which has led to the development of highly agile missiles such as the IRIS-T and ASRAAM which can be targeted at off-boresight angles that are greater than their sensor's gimbal limits. Exploiting this capability requires a helmet-mounted sight to be used to detect the pilot's line of sight (in the direction of the target aircraft) and to feed these data into the guidance computer in the missile. Post launch, the missile will manoeuvre until the target is within its sensor's field of view. This gives modern short-range missiles a LOAL capability.

Radar-guided missiles can operate in both LOBL and LOAL modes. In a LOBL mode, the aircraft's radar is used to detect a target and provide data to the missile that enables it to acquire the target prior to launch. Post launch, the missile may receive target position updates through a data link. Radar-guided missiles may also contain a navigation system to enable a LOAL mode to be employed. Here, the aircraft provides the missile with target co-ordinates. As the target may be travelling at a high speed and manoeuvring, a post-launch data link is used to update the missile with the target's current position. The missile will turn on its radar sensor late in the engagement and guide itself to the target. This method of operation enables a relatively stealthy attack to be prosecuted, thereby providing a very high probability of kill (P_k).

This is an important consideration in any air-to-air engagement, and therefore it is common for the Weapon Aiming Computer to continuously generate a prediction of P_k and display this to the pilot as a Launch Success Zone. To achieve this, the Weapon Aiming Computer employs a high fidelity missile fly-out model based on the real performance of the missile for the given launch conditions and the relative range and bearing to the target. Fire cues will also be given to the pilot in order to maximise the P_k. As the fly-out model could, in its purest form, be a 6-DOF accurate representation of the true performance of the missile, then the computer processing power required to provide a real-time prediction of P_k can be very demanding. As the window for a successful prosecution of an attack can depend on split-second timing, the overall P_k of the system depends not only on the performance of the missile, but also on that of the aircraft sensor and computing systems. The integration of the components of an air-to-air engagement system is therefore a complex task but one which is essential in providing a leading-edge capability.

5.6.3   Air-to-Air Weapons Training

Training for the use of Air-to-Air weapons primarily covers short-range missiles with emphasis placed on the tactics and flying skills required for a LOBL engagement in a dog-fight situation. Clearly, during training, it is impractical to fire a live missile at an opposing aircraft. Therefore, special missile variants are built which contain only the target acquisition sensor and associated electronics. This enables aircrew to undertake realistic training in dog-fight situations, tracking the target and locking on without the need to risk firing a live missile. These 'acquisition' rounds can generally be used without any aircraft system changes in place of the live missiles. Therefore, the displays, controls and functionality implemented for the operational missile can be used in training. Should an image of the pilots view through the head-up display (HUD) be recorded, this can be used to aid in post-sortie debriefs to determine the success on the target engagement.

This is a cost-effective method of providing a training capability, but even with HUD recordings, there is inadequate information available to discuss how the engagement scenario developed and how it could have been more successful.

Experience from the Vietnam War identified the need for more realistic training where in an instant, the aircraft prosecuting an attack could itself become the target. This led to the development of instrumentation which, in conjunction with an acquisition missile, could be used to record the manoeuvring of the aircraft and any critical actions such as target acquisition, lock-on and pilot demands to fire the missile.

Maintaining the ethos of minimising the impact on the launch aircraft systems, the instrumentation (referred to Air Combat Manoeuvring Instrumentation – ACMI) was implemented in a package similar to a short-range Air-to-Air missile body tube. This reduced the aeromechanical integration issues. The level of systems integration of the ACMI pod could be varied to suit the needs of the aircraft. In the simplest implementation, the pod would employ power from the aircraft and pick up on missile sensor pointing commands, lock-on and fire discrete signals. To capture aircraft manoeuvres, the pod also contained a basic inertial measurement system. All the data captured by the pod would be transmitted to the ground over an RF data link. For more complex scenarios, aircraft manoeuvring data could be transmitted to the pod over a MIL-STD-1553B data bus.

The need to downlink data meant that the pods (and therefore the training) had to be operated on a specially instrumented range. Several such ranges were built around the world for use by the United States and its allies. These ranges were equipped with data replay and de-briefing facilities such that training sorties could be critically assessed.

Whilst the ACMI system provided an effective training system, it meant that pilots needed to train wherever the range was located. Although the United Kingdom did operate a range in the North Sea, the inconvenience of having to take pilots, aircrew and the necessary logistics support for the aircraft to the ranges proved to be expensive.

The advent of high-density recording media means that the ACMI principle has been extended such that captured data can be time-stamped, using a common time base such as GPS time and recorded in the pod for replay on the ground. The use of a common time-stamp means that multiple aircraft can train and all their data replayed on the ground in complete synchronisation. The added benefit is that aircraft no longer need to use a dedicated ACMI range, giving greater flexibility to cost-effectively train aircrew wherever they are located.

For target engagements using medium-range Air-to-Air missiles, the aircrew interactions with the aircraft systems will primarily be based on the use of radar modes to detect and classify targets at longer ranges. A training variant of the live missile may be employed to provide the correct system interactions with the aircraft, but as for short-range Air-to-Air missiles, the weapon will only simulate the correct responses up to the point of firing.

As modern air combat becomes more complex with more potential targets and friendly forces intermingled, the need for a new, comprehensive training system is demanded.

As more aircraft are updated to include tactical data links such as Link-16, then a greater awareness of the battlespace becomes available to the crew. However, with the exception of major exercises such as 'Red Flag' (a complex battle scenario for NATO air forces where many airborne assets are operated in opposing forces, interacting with ground defences, early warning aircraft, etc.), cost-effective training can be prohibitively expensive. Training systems are being introduced whereby many targets can be simulated and displayed to aircrew and, using special modes, can be interrogated and classified as though using real information from subsystems such as the radar and transponder. Coupled with the use of simulated or training weapons, the use of Link-16 can also enable simulated battlespace data to be up-linked from the ground to create a complex battlespace for crew training. This level of connectivity can also be used to add ground-based aircraft simulators into the battlespace so that more aircrew can participate simultaneously in the training without the need to have large numbers of aircraft in the sky. Such an immersive training system can provide an unparalleled opportunity to train without the need to commit complex airborne assets. On-board RF data links can be used to route aircraft data back to the ground for post-sortie de-briefing.

The implementation of such a capability into the aircraft system employs existing training modes supplemented by RF data links. As the operational system will probably already include the necessary level of integration to enable real targets in a complex battlespace to be engaged, the main addition required to implement the training capability is an investment in the ground-based simulation capability.

Whilst simple training systems will always have a place in the development of aircrew skills, it is likely that in the coming years a greater emphasis will be placed on immersive training using synthetic environments within the operational cockpit.

Further Reading

Society of Automotive Engineers. (2005) Common Launch Acceptability Region Truth Data Generator Interface Control Document for the CLAR Approach. *Aerospace Information Report 5788*, Society of Automotive Engineers, Warrendale.

Society of Automotive Engineers. (2007) Common Launch Acceptability Region Approach Interface Control Document. *Aerospace Information Report 5682*, Society of Automotive Engineers, Warrendale.

Society of Automotive Engineers. (2008) Common Launch Acceptability Region Approach Rationale Document. *Aerospace Information Report 5712*, Society of Automotive Engineers, Warrendale.

United States Department of Defense. (2000) *World Geodetic System 1984*, National Imagery Mapping Service, Bethesda (published as NIMA TR8350.2, 3rd edn, Amendment 1).

6

Weapon Interface Standards

6.1 Chapter Summary

From the earliest days of guided weapons, weapon designers have defined their interfaces to optimise the requirements against technologies that have been available. This has led to a plethora of different interfacing systems existing on aircraft. Standards such as MIL-STD-1760 have sought with some success to reduce the number and variation of interfaces and thereby improve interoperability of different weapons across many platforms. With the success of interfacing standards, the weapons integration community has also developed standards that ease other areas of integration such as protocol standards that facilitate greater interoperability in aircraft computing systems. This chapter will introduce standardisation and its role in improving interoperability and reducing integration costs and timescales before describing the MIL-STD-1760 Aircraft/Store Electrical Interconnection System (AEIS). Chapter 7 will provide an overview of other relevant standards.

6.2 Benefits of Standardisation

As weapon designers of early guided weapons defined their interfaces to optimise the requirements using the available technology, an aircraft would have to provide a number of different interfaces to support the weapons in its inventory. From the aircraft designer's viewpoint, there was little opportunity to re-use existing interfaces or functionality. The answer to this was a drive, initially by the US DoD, to introduce a greater level of standardisation. Standardisation brings real benefits, not least of which is the opportunity to benefit from previous integration programme investments and to improve interoperability, thereby reducing the cost of integration programmes. However, it is recognised that not everything can be standardised.

Aircraft Systems Integration of Air-Launched Weapons, First Edition. Keith A. Rigby.
© 2013 John Wiley & Sons, Ltd. Published 2013 by John Wiley & Sons, Ltd.

The overriding need of any integration programme is to achieve the required system performance. This means that it is not possible to standardise every aspect of the design of aircraft and weapons purely to simplify integration. For example, it is unrealistic to require the systems architecture of all aircraft to be identical such that a single modification could be developed and applied across all aircraft.

From the systems integration viewpoint, the first effort to standardise was the development of the MIL-STD-1760 AEIS in the United States. This overcame the need for aircraft to provide numerous interface connectors to cater for different weapons. The standard also provisions a large signal set for use by stores, thereby fostering a greater level of interoperability.

Standardising the location of the connectors on the store also enables aircraft to provision standard hard-points for retraction mechanisms, resulting in a simplified pylon structural design. Such requirements are defined by MIL-STD-8591.

From an integration viewpoint, there are many activities that must be performed to modify the aircraft, demonstrate safe carriage and release and to ultimately provide a working and safe system to the customer. Of all the activities, it is not uncommon for the systems integration activities alone to exceed 40% of the total integration costs. Once an aircraft has fully functioning standard interfaces, then the systems upgrades required to integrate further new weapons can be achieved purely by software upgrades (albeit, these upgrades can still be significant). On existing programmes, this has demonstrated significant benefits. Studies undertaken by the NATO Industrial Advisory Group (NIAG) have also estimated that by adopting standard interfaces, the cost of the required aircraft software modifications could be between 15% and 40% of the equivalent systems integration costs associated with weapons with a bespoke interface. It is therefore essential that to reduce the cost of weapons integration programmes, a greater use of standardisation is required.

Whilst the adoption of formal technical standards brings benefits, it is also possible to simplify the integration activity by re-using existing system interfaces and functionality. For example, the AIM-9 Sidewinder analogue interface has become a de facto standard for short-range air-to-air missiles. Missiles such as the AIM-9X, ASRAAM and IRIS-T are required to operate from the standard Sidewinder analogue interface with operation from a digital interface being optional. This is a way of improving interoperability and also of reducing initial integration costs. However, it must be stressed that this is not an optimum solution as legacy interfaces are unlikely to be able to exploit the full capabilities of modern weapons (e.g. high off-boresight target acquisition of airborne targets by short-range missiles) employing a digital interface.

To date, there has been little or no information of the potential savings which can be achieved by the adoption of technical standards. There may well be an initial expense in adopting standardisation, but the real benefits come later in the aircraft life cycle when further weapons are integrated. In the United States, some cost comparisons of integrating a MIL-STD-1760-compliant weapon to an aircraft which already has the capability embodied have been performed. In unpublished work, it has been estimated that over 40% of programme costs can be saved largely in logistics, new hardware and software changes.

6.3   MIL-STD-1760 AEIS

The standardisation activities which resulted in MIL-STD-1760 commenced in the early 1980s as a means of reducing the proliferation of multiple interfaces. Consequently, the standard has tried to encompass every facility that the weapon integrator could require in a standard interface. Whilst MIL-STD-1760 is by far the most commonly used interface by smart weapons, it does have its limitations, primarily around the size of the connector and the break force during a stores release. With the trend for smaller, lighter and more capable systems, the large signal set is also considered to provide more capability options than necessary for the future.

The adoption of a relatively large connector by the standard means that connector real estate and weapon body strengthening purely to withstand the connector break force becomes a significant constraint on the design of smaller weapons. Therefore, in 1998, the US Air Force enlisted the help of the SAE to develop a new interface standard for weapons in the sub-125 kg class, requiring a reduced signal set and a very low connector break force. The result of this effort was AS5725 (Interface Standard, Miniature Mission Store Interface).

With the advent of small weaponised unmanned platforms, new requirements emerged for an even smaller class of munition (in the sub-25 kg class). In response to this, the US Navy approached the SAE to develop a new standard which had a minimised signal set and provided for different interconnect technologies to be adopted, dependent of the actual munition size and technology. The result of this work was AS5726 (Interface Standard, Interface for Micro Munitions).

A key part of MIL-STD-1760 is the protocols employed for safe control of the store. Much effort was expended in developing these to the point where certification bodies could accept that safe control could be maintained. Therefore, when developing the Miniature Mission Store and Micro Munition interfaces, these protocols were retained, albeit using a different data link layer and physical layer.

When considering the three interfaces, it should be noted that whilst there is a progressive reduction in the signal set and the adoption of different technologies, the functional capability provided by all is similar. However, before looking at the recent weapon interfacing standards, there is a need to gain an understanding of MIL-STD-1760.

6.3.1   MIL-STD-1760 Interface Points

Figure 6.1 shows the interface points in the integrated weapon system. From the aircraft integrator's viewpoint, the main interface is the Aircraft Station Interface (ASI). This interface can be located anywhere in the aircraft pylon. The associated connector on the store is the Mission Store Interface (MSI – Mission Store is used as the generic description for any store that is fitted to the aircraft for a specific mission; e.g. a GPS-guided bomb). An umbilical cable is used to connect the ASI to the MSI, and whilst the standard includes requirements for such a cable, its length is not defined, leaving this to the aircraft designer to determine. To enable both aircraft and stores to be independently designed, the standard defines the interface requirements at both the ASI and MSI.

MIL-STD-1760 also makes provision for multi-store carriers called Carriage Stores which enables the platform's total weapon load-out to be extended. The standard therefore identifies two further interfaces as shown in the figure: the Carriage Store Interface (CSI) and the Carriage Store Station Interface (CSSI).

6.3.2   Connectors

MIL-STD-1760 defines two signal sets, and these are allocated to two separate connectors. The signal sets are referred to as the primary interface and the auxiliary interface. Whilst the two connectors use identical shell size connectors, the inserts are very different to account for the different signal sets. Figure 6.2 shows these two insert formats.

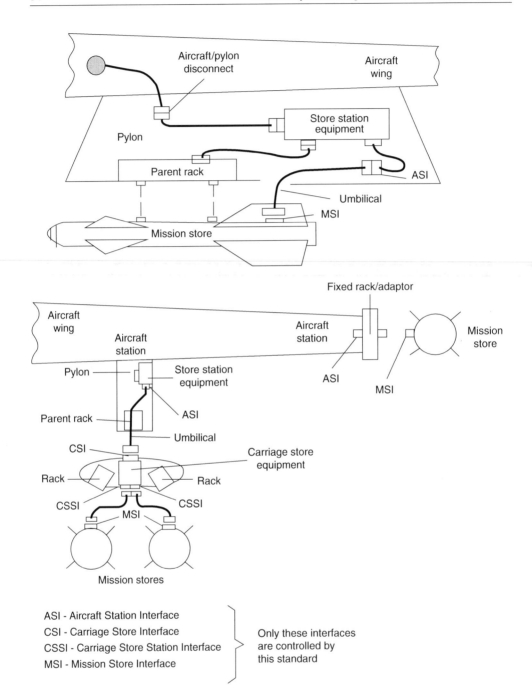

Figure 6.1   MIL-STD-1760 interface points (from MIL-STD-1760E)

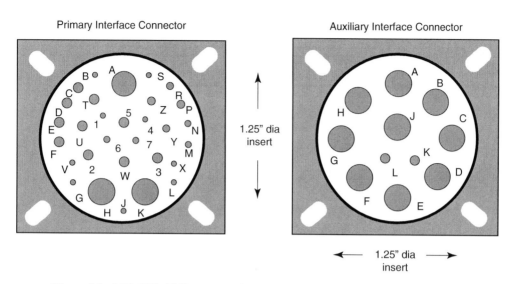

Figure 6.2   MIL-STD-1760 connector insert arrangements (from MIL-HDBK-1760A)

6.3.3   Signal Sets

When MIL-STD-1760 was originally defined, the intent was to provision an interface which could replace all previous store interfaces, introducing standardisation for future stores. The primary signal set contains all the possible services that were envisaged, including power supplies, analogue signals, a number of discrete signals and a dual MIL-STD-1553B data bus. It was also acknowledged that stores could use only the signals required to implement an intelligent interface with the aircraft. Also, it was acknowledged that aircraft may not need to provide all signals defined by the standard. Therefore, interface classes were introduced to maintain a level of standardisation. For aircraft to be compliant, either a Class I (the full signal set) or a Class II (reduced signal set) interface must be provided. The auxiliary signal set was defined to provide additional high-power supplies. Table 6.1 and Table 6.2 identify the two signal sets, the characteristics of each signal and the associated interface classes.

6.3.4   GPS RF Signal Distribution

When loaded to an aircraft, a weapon's GPS antenna may be obscured (e.g. at an under-fuselage station or when loaded in an internal weapons bay). Therefore, for weapons that are required to lock to GPS satellites prior to launch, there is a need to provide a feed of the RF signal. The High Bandwidth 1 (HB1) Type B interface has been included in the MIL-STD-1760 interface to provide this facility. However, as identified in Chapter 4, the routing path through the aircraft for these sub-noise signals will vary from platform to platform. The standard therefore introduces requirements aimed at defining the quality of the RF distribution network which the aircraft must provide. It is here where the standard deviates from its own standardisation points identified in Figure 6.1. Instead of defining the signal requirements at the ASI, the interface is actually defined at the lower umbilical connector (the mating half to the MSI). The standard can therefore treat the aircraft system as a two-port network, with the

Table 6.1   MIL-STD-1760 primary signal set

Primary interface signal set		
Signal	Characteristics	Notes
Interlock	Nominal 28 V dc low current (100 mA max) discrete signal	Enables the aircraft to determine the mated status of the store connector
Interlock Return	Low current (max 100 mA) return line	Provides a return signal for the Interlock line
28 V DC Power 1	28 V, 10 A dc power supply. Current/time profile defined by the standard	Store 'logic' supply for powering non-safety-critical functions. Must be independently controllable from other supplies
Power 1 Return	10 A rated return line for 28VDC1	Separate returns required for all power supplies
115 V/200 V AC 3-Phase Power	10 A per phase, three-phase power supply. Standard defines current/time profile, phase rotation, phase imbalance, power factor and so on	Aircraft may energise power at any time under the assumptions that store functions powered are either not safety-critical or sufficient interlock exists such that store safety is not significantly degraded. Aircraft must not energise power unless connector is mated (reference the Interlock/Interlock return signals)
115 V/200 V AC neutral	Neutral connection for AC power	Reference connection to carry phase imbalance currents
270 V DC Power	270 V, 10 A dc power supply. Current/time profile defined by the standard	Aircraft may energise power at any time under the assumptions that store functions powered are either not safety-critical or sufficient interlock exists such that store safety is not significantly degraded. Aircraft must not energise power unless connector is mated (reference the Interlock/Interlock return signals)
270 V DC Return	10 A rated return line for 270 V DC Power	Separate returns required for all power supplies
28 V DC Power 2	28 V, 10 A dc power supply. Current/time profile defined by the standard	Store supply for powering safety-critical functions. Must be independently controllable from other supplies
Power 2 Return	10 A rated return line for 28VDC2	Separate returns required for all power supplies
Mux A	MIL-STD-1553B data bus	Time division, command/response multiplex data bus. Voltage levels and zero-crossing constraints of signals defined at the ASI
Mux B	MIL-STD-1553B data bus	See 'Mux A'
Address Lines (A0, A1, A2, A3, A4, Parity and Return)f	Low current dc discrete. Voltage range between 3.5 and 31.5 V	Used by the store to determine the Mux A/Mux B RT address. Intention is that the address lines and Parity line are connected to the Address Return line, such that an odd number of logic '1' states are set (i.e. not connected to the Return interface). Stores may also monitor this interface to determine the mated status of the connector

(continued)

Table 6.1 (*continued*)

Primary interface signal set		
Signal	Characteristics	Notes
Up Fibre Channel (Class I interface only)	Fibre Channel copper interface with a data transmission rate of 1.0625 Gbaud. Up Fibre Channel is the store to aircraft path	A high-data rate communications path defined by AS5653 for the transfer of digital video, digital audio or data messages replicating the MIL-STD-1760 and/or MIL-STD-1553B data bus protocols
Down Fibre Channel (Class I interface only)	Fibre Channel copper interface with a data transmission rate of 1.0625 Gbaud. Down Fibre Channel is the aircraft to store path	See 'Up Fibre Channel'
High Bandwidth 1	Type A: 20 Hz–20 MHz. Type B: 20 MHz–1.6 GHz. 50 Ω impedance	Primarily used for the transfer of timing pulses (Type A signals) or GPS RF signals (Type B)
High Bandwidth 3	Type A: 20 Hz–20 MHz. 75 Ω impedance	Primarily used for the transfer of analogue video signals
Low Bandwidth	300 Hz–3.4 kHz audio signals. 600 Ω impedance	Primarily used for the transfer of analogue audio signals
Release Consent	Nominal 28 V dc low current (100 mA max) discrete signal. Return is via the 28VDC2 power return line	Provided to satisfy a safety function with consent only being enabled when the aircraft determines that safety criteria have been met
Structure Ground	0.2 V maximum voltage drop when conducting a 10 A current. Must be capable of withstanding the over-currents defined for the primary interface power supplies	Electrical connection provided to minimise electric shock hazard for personnel. Must not be used as either a signal return or power return except under fault conditions within the store
Fibre Optic 1 (Class I interface only)	Not used	Originally provisioned for a fibre optical interface

signal inputs being defined by the GPS system specification (e.g. signal power levels available at the aircraft antenna). The standard defines requirements for Voltage Standing Wave Ratio, effective gain, group delay, Figure of Merit and signal path dc offset at the interface. The intention of the standard is that any aircraft system that meets the stated requirements will provide sufficient signal fidelity in terms of power and SNR for Mission Stores to acquire and track the GPS signals.

The standard recognises that the aircraft system will need to boost signal levels to account for losses through cables and connectors, that themselves will introduce additional noise into the system. Amplification, whilst boosting the wanted signals, will also boost the noise. The increased noise can become problematic, leading to increased acquisition times for the store's GPS receiver. Whilst the routing network signal gain can be set to counteract the signal losses

Table 6.2 MIL-STD-1760 auxiliary signal set

Auxiliary interface signal set		
Interlock	Nominal 28 V dc low current (100 mA max) discrete signal	Enables the aircraft to determine the mated status of the store connector
Interlock Return	Low current (max 100 mA) return line	Provides a return signal for the interlock line
Aux 28 V DC	28 V, 30 A dc power supply. Current/time profile defined bythe standard	Store 'logic' supply for powering non-safety-critical functions. Must be independently controllable from other supplies
Aux 28 V DC Return	30 A rated return line for Aux 28 V DC	Separate returns required for all power supplies
Aux 115 V/200 V AC 3-Phase Power	30 A per phase, three-phase power supply. Standard defines current/time profile, phase rotation, phase imbalance, power factor and so on	Aircraft may energise power at any time under the assumptions that store functions powered are either not safety-critical or sufficient interlock exists such that store safety is not significantly degraded. Aircraft must not energise power unless connector is mated (reference the Interlock/Interlock return signals)
Aux 115 V/200 V AC Neutral	Neutral connection for AC power	Reference connection to carry phase imbalance currents
Aux 270 V DC Power	270 V, 30 A dc power supply. Current/time profile defined by the standard	Store 'logic' supply for powering non-safety-critical functions. Must be independently controllable from other supplies
Aux 270 V DC Return	30 A rated return line for 270 V DC Power	Separate returns required for all power supplies
Structure Ground	0.2 V maximum voltage drop when conducting a 30 A current. Must be capable of withstanding the over-currents defined for the auxiliary interface power supplies	Electrical connection provided to minimise electric shock hazard for personnel. Must not be used as either a signal return or power return except under fault conditions within the store

in cables and connectors, the management of noise within the system is a key factor. Figure of Merit (defined as $G_e - 10\log_{10} T$, where G_e is the effective gain in decibels and T is the output noise temperature measured in kelvins) is used as a measure of the quality of the routing network, and the standard identifies levels for each of the GPS signals that, if met, will ensure a store complying with the standard's SNR requirements can function.

When calculating the Figure of Merit, it is necessary to understand that each component in the routing path will introduce its own noise that will be added to the noise at its input. The total noise is boosted by the effective gain (or loss) of the component. The complete distribution system from GPS antenna to the MSI end of the umbilical cable can be modelled with each piece of cable, connector, amplifier and so on being a separate component (see Figure 6.3).

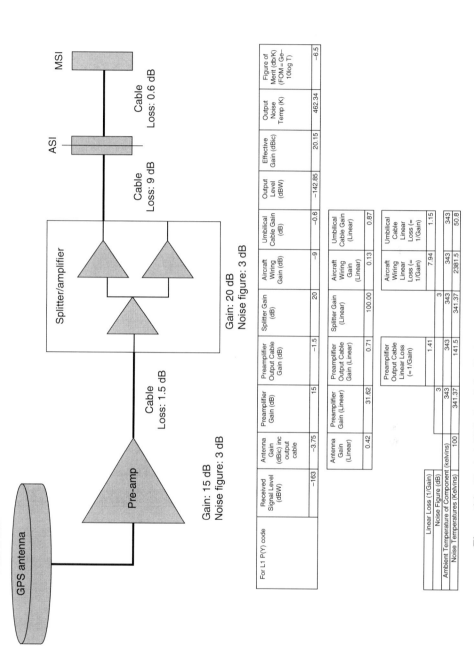

Figure 6.3 Typical GPS signal distribution model and Figure of Merit calculation

For each component, it will be necessary to calculate the noise temperature. For a component such as an amplifier, there will usually be a Noise Figure defined in its specification. Noise figures can be converted to an equivalent noise temperature using the following formula:

$$t_n = (F - 1) T_0$$

where F is the specification Noise Figure and T_0 is the ambient temperature where the component is located. (For components mounted inside the airframe, it will be necessary to estimate an appropriate temperature. In Figure 6.3, a temperature of 343 K (70°C) is assumed.)

For a component such as a section of RF cable, the noise temperature is calculated using the following formula:

$$t_n = \left(\left(\frac{1}{g} \right) - 1 \right) T_0$$

where g is the anti-log of the gain expressed in decibels (dB) (where cable loss is expressed as a negative decibel gain) and T_0 is, again, the ambient temperature where the component is located.

Once the noise temperature for each component has been calculated, it is possible to determine the overall noise temperature for the distribution system using the formula:

$$\text{System Noise Temperature} = T_{antena} + \left(t_1 + \left(\frac{t_2}{g_1} \right) + \left(\frac{t_3}{g_1 g_2} \right) + \left(\frac{t_4}{g_1 g_2 g_3} \right) + \cdots \right.$$
$$\left. + \left(\frac{t_n}{g_1 g_2 g_3 \cdots g_n} \right) \right)$$

where, in this example, $T_{antenna}$ is the noise temperature of the antenna (assumed to be 100 K), $t_1 \ldots t_n$ is the noise temperature of the individual component and $g_1 \ldots g_n$ is the linear gain (or loss) of each component.

Comparing the resulting performance of this example system with the requirements of the standard shows that the Effective Gain is in the middle of the gain range (+9 dBic to +30.5 dBic) with Figure of Merit very good (much greater than the −35 dBK^{-1} required for the L1 P(Y) signal path (see Chapter 4 for an explanation of L1 P(Y))). Therefore, the example aircraft system should provide an acceptable RF distribution network for the GPS signals. As a general rule-of-thumb, low-noise amplifiers and short, high-quality cables should be used wherever possible in the front end of the distribution system to avoid excessive signal degradation.

The standard also recognises that some GPS receivers provide a dc bias for use as an antenna power supply. Although controlled by the standard, the RF distribution network should be designed to work when this dc offset is present, although the standard does control the maximum offset.

6.3.5   Data Protocols

The MIL-STD-1553B data bus at the heart of the interface is provisioned for the control of stores. The data bus operates in a dual standby redundant mode. That is, there are two data paths (MUX A and MUX B) where only one path is used at any one time. Should data bus communications fail on one bus (e.g. MUX A), the Bus Controller will attempt to utilise the second path (MUX B) to transfer the data.

The data bus control includes non-safety involved aspects of store control such as store discovery and Transfer Alignment, but also enables the control of high-integrity functions such as firing of missile rocket motors.

MIL-STD-1760 adopts the basic protocols of MIL-STD-1553B but defines a series of standard messages to effect store control and monitoring. These are the Store Description Message (SDM – primarily used to enable the aircraft to identify the store type loaded at a particular station), the Store Control Message (SCM – used to control the state of a store, particularly safety-critical commands) and the Store Monitor Message (SMM – primarily used as a status message reflecting the safety-critical condition of the store). An additional standard message (Aircraft Description) is used should the store have a need to know from which aircraft it is being operated. The standard also provisions a method for the transfer of large data files between the aircraft and the store (Mass Data Transfer – MDT, see Section 6.3.8). The WDO may also use spare capacity on the data bus for 'User-Defined' messages.

A prerequisite of enabling control of safety-critical functions is that a high level of integrity must be assured. MIL-STD-1760 therefore provisions a number of protocol elements which act together to satisfy this requirement. Each of the protocol elements will be discussed in the following paragraphs.

The standard requires that each message must include thirty 16-bit words regardless of whether there is data to be transmitted (where there is no data to be transmitted, unused words are filled with zeros). Each standard message has its own 16-bit unique header word (word 1 in the message). For the SCM and SMM messages, words 2 and 3 are Invalidity Words which are used to identify if any of the 30 data words in the message are deemed by the subsystem compiling the message to contain incorrect or erroneous data.

Whilst the SCM and SMM contain data words for non-safety-critical control, the important words for safety-critical control are the Critical Control and Critical Authority word sets (SCM words 4, 5, 6 and 7) and the Critical Monitor messages (SMM words 4 and 5).

The Critical Control 1 word (SCM word 4) format is shown in Table 6.3. It is this word which provides the platform's demands for the control of weapon pyrotechnics.

For a safety-critical store function to be demanded, the relevant bits (D_3–D_{10}) must be set to a logic '1', the Identifier field must have the correct bit pattern set and the Address Confirm field set to match the station's RT address. If either the Identifier or Address Confirm field is set incorrectly, then the store must reject the message. The Critical Control 2 word (SCM word 6) has different functions allocated to bits D_3–D_{10}, but the logic behind selecting the appropriate function is as detailed earlier.

Each Critical Control word has an associated 16-bit Critical Authority word (SCM words 5 and 7). The sixteen bits are set depending on the selection of bits D_3–D_{10} in the associated Critical Control word and based on a Bose–Chaudhuri–Hocquenghem 31, 16, 3 polynomial. Table 6.4 shows the format of the Critical Authority words.

The Critical Monitor words contained in the SMM are used by the store to indicate the store state in response to actions demanded by the Critical Control words. The format of the Critical Monitor 1 word is shown in Table 6.5. It should be noted that the Critical Monitor 2 word is of a similar format but relates to the functions selected by the Critical Control 2 word.

When the aircraft requires a safety-related state change to be made in the store, the SCM is transmitted and then the SMM is interrogated to check that the demanded actions are being

Table 6.3   Critical control 1 word format

Field name	Bit number	Description
Store Control (note 1)	00	D_{10} = Fire, Launch or Release
	01	D_9 = Jettison
	02	D_8 = Commit to Separate Store or Submunition
	03	D_7 = Execute Arming
	04	D_6 = Preset Arming
	05	D_5 = Select Store
	06	D_4 = Initiate Interruptive BIT
	07	D_3 = Release/Launch Mode (see note 2)
Identifier	08	D_2 (see note 3)
	09	D_1 (see note 3)
	10	D_0 (see note 3)
Address Confirm	11	A_4 (see note 4)
	12	A_3 (see note 4)
	13	A_2 (see note 4)
	14	A_1 (see note 4)
	15	A_0 (see note 4)

Note 1: When a data bit is set to logic '0', the function is required to be inactive. When a data bit is set to logic '1', the function is required to be active. Data bits reset to logic '0' require the function to be deactivated.

Note 2: When bit D_3 is set to logic '0', the Mission Store is released from S&RE. When bit D_3 is set to logic '1', the Mission Store is launched by forward firing.

Note 3: The IDENTIFIER field shall be set as follows:

D_2	D_1	D0	
0	0	0	Reserved
0	0	1	Mission Store
0	1	0	Carriage Store
0	1	1	Reserved
	thru		
1	1	1	Reserved

Note 4: Bits A_0–A_4 are set to match the interface RT address discrete lines.

undertaken. Should the store not respond with a correct SMM, then the aircraft can disable the store to ensure overall integrity is maintained. The level of integrity provided by the use of the Header and Invalidity words, the Critical Control format, coupled with the Critical Authority word and the monitoring provided by the Critical Monitor format, is extremely high. This means that it is improbable that a correctly formatted and sequenced data exchange can be erroneously initiated. However, recognising that all data is likely to be generated by software, the safety authorities will insist that a level of electrical control is also included as a part of authorising a safety-related state change in the store. The Release Consent discrete signal is provided for this purpose with the standard mandating that the store cannot action a correctly formatted SCM unless the Release Consent signal has been in the enabled state for greater

Table 6.4   Critical authority word format

Field name	Bit number	Description
Coded Check	00	$C_{14} = D_{10} + D_9 + D_6 + D_1 + D_0$
	01	$C_{13} = D_9 + D_8 + D_5 + D_0$
	02	$C_{12} = D_8 + D_7 + D_4$
	03	$C_{11} = D_7 + D_6 + D_3$
	04	$C_{10} = D_{10} + D_9 + D_5 + D_2 + D_1 + D_0$
	05	$C_9 = D_{10} + D_8 + D_6 + D_4$
	06	$C_8 = D_{10} + D_7 + D_6 + D_5 + D_3 + D_1 + D_0$
	07	$C_7 = D_{10} + D_5 + D_4 + D_2 + D_1$
	08	$C_6 = D_{10} + D_6 + D_4 + D_3$
	09	$C_5 = D_9 + D_5 + D_3 + D_2$
	10	$C_4 = D_{10} + D_9 + D_8 + D_6 + D_4 + D_2 + D_0$
	11	$C_3 = D_9 + D_8 + D_7 + D_5 + D_3 + D_1$
	12	$C_2 = D_{10} + D_9 + D_8 + D_7 + D_4 + D_2 + D_1$
	13	$C_1 = D_{10} + D_8 + D_7 + D_3$
	14	$C_0 = D_{10} + D_7 + D_2 + D_1 + D_0$
Reserved	15	Reserved (shall be set to logic '0')

Table 6.5   Critical Monitor 1 word format

Field name	Bit number	Description
Store State	00	Fired, Launched or Released
(note 1)	01	Jettisoned
	02	Committed to Store or Submunition Separation
	03	Armed
	04	Arming Preset
	05	Store Selected
	06	Store in Interruptive BIT
	07	Store to be Released/Launched (see note 3)
Demanded State	08	D_{10}
(note 2)	09	D_9
	10	D_8
	11	D_7
	12	D_6
	13	D_5
	14	D_4
	15	D_3

Note 1: Data bits 00–07 when set to logic '1' indicate that the associated store state is true.
Note 2: The demanded state is a monitor of the last received demanded state in the associated critical control word.
Note 3: The released/launched field is set as follows:
D_3 is set to logic '0' indicates the Mission Store is to be released from S&RE.
D_3 is set to logic '1' indicates the Mission Store is to be launched by forward firing.

than 10 ms. Should the aircraft subsequently reset the Release Consent signal, the store must reject further SCMs within 10 ms of the inhibit state being detected.

6.3.6   Data Entities

MIL-STD-1760 was instigated to introduce greater interoperability between aircraft and stores. Whilst a number of standard messages are defined, MIL-STD-1760 allows the WDO to define messages to meet their own requirements for data transfer. This in itself could work against the interoperability principle. However, to reduce this, the standard defines a number of data entities (almost 200) which the WDO is strongly encouraged to use. These cover almost everything that the WDO (or the ADO) could require to enable store initialisation, priming and control. Examples include True Airspeed, Angle of Attack, Barometric Altitude and Aircraft Roll Angle to name but four. Parameters such as these use linear data entities such as velocity, angle and distance, and the standard also defines how these should be implemented in terms of format (unsigned, scientific, 2's complement, etc.) and values for the most and least significant bits. Table 6.6 gives an example.

6.3.7   Time Tagging

Whenever data is transferred from the aircraft to the store, there is a finite time taken for data to be sensed, processed by the aircraft subsystems, transmitted over the data interface and read into the store's memory ready for processing. The aircraft system architecture can add significant delay to the time of validity of a piece of data, particularly if the data has to traverse a number of data buses (e.g. to be transmitted from a Mission Computer across the Avionics Data Bus to the SMP, and then from the SMP across the Weapons Data Bus to the Store's RT). As the exact delay introduced will be different for different aircraft, there is a need for the weapon to determine when the data was valid (i.e. the time at when the data was sensed). MIL-STD-1760 provides a protocol for the time-tagging of data which is critical to the operation of the store (e.g. aircraft position and velocity data).

Table 6.6   Data entity example

Data entity	Entity type	Entity description	
True Airspeed MSP & LSP	Velocity (M) & Velocity (L)	Dual-precision data entity. True airspeed of the aircraft, represented as positive when the aircraft is travelling through static air in the X_a direction (see Figure 5.1)	
Word type	Format	MSB value	LSB value
Velocity (M) in metres/second	2's complement	$-(2^{13}) = -8192$	$2^{-2} = 0.25$
Velocity (L) in metres/second	Unsigned	$2^{-3} = 0.125$	$2^{-18} = 3.8 \times 10^{-6}$

The aircraft system may use time for many reasons. This could be real time (based on UTC) or could be system time measured from the time when the systems were initialised. A store may use UTC or have its own internal system time that is different from that used by the aircraft. Therefore, if the aircraft tags data at the time it was produced, this will be meaningless to the store unless the store system understands the difference between its internal time and the aircraft's system time. When the store is initialised, there will therefore be a need for the aircraft to synchronise time across the aircraft and store systems.

MIL-STD-1553B provides a number of 'Mode Codes' which are used to control the RTs connected to the data bus. One such Mode Code is the Synchronise with Data Word Mode Code. Here, the BC will transmit a Mode Code with an associated data word which contains the current aircraft system time to the store RT. The store system can then synchronise its internal clock to the aircraft clock.

Some aircraft systems do not require absolute time and so the system clock may have a maximum value before it resets to zero. The aircraft must therefore transmit data to the store which identifies this maximum time at reset. The maximum time that the clock can achieve before it resets must also be longer than the maximum data latency that the store can accept without degrading its capability.

The data transmitted with the Synchronise with Data Word Mode Code enable the store system to synchronise its internal time with the aircraft's system time. When a piece of time-tagged data is received by the store, it can then resolve the time the data was valid at the point it was sensed by the aircraft in terms of its own internal time. This enables the store to know when the aircraft data were valid and to use them accordingly. However, in some cases, the store may determine that the time that the aircraft data were sensed appears to be later than the time when it was received by the store (due to the aircraft system time having been reset). It is for this reason that the store needs to know the maximum time that the aircraft clock will reach before it is reset, thereby enabling in cases such as this, data latency to be resolved. The actual data latency that can occur must be recorded in the aircraft/store ICD.

6.3.8   Mass Data Transfer

Some stores require the transfer of large data files. For example, a cruise missile may contain a map data base, detailed target data to enable scene matching during the terminal phase and so on. Such files can be several gigabytes in size. If these were to be transferred over the MIL-STD-1553B data bus using unused time slots, the time taken to download the data would be operationally unacceptable. In Section 6.3.5, it was noted that MIL-STD-1760 provides a facility (MDT) to transfer large data files between the aircraft and store (or from the store to the aircraft). When an MDT is initiated, all other transactions on the data bus are suspended until the data transfer is complete. This makes the total usable data bandwidth available solely for MDT, therefore, speeding up the time to transfer a large file.

MDT uses three standard messages: Transfer Control (TC), Transfer Monitor (TM) and Transfer Data (TD). The TC message is an eight-word message used by the aircraft to set up the required data transfer mode (e.g. select data download (transfer data from the aircraft to the store), data upload (transfer data from the store to the aircraft), erase functions, etc.). The TM message is an eight-word message set by the store to indicate the status of the MDT as demanded by the TC message. The TD message is a 30-word message used to convey the actual data either to be downloaded or uploaded.

MDT provides a very flexible, albeit complex capability for transferring large data files. The reader is recommended to reference MIL-HDBK-1760A which includes around 175 pages of guidance and examples on the subject in much greater detail than could be justified in this book.

6.3.9 High-Speed 1760

Whilst MDT provides the capability to transfer large data files, the MIL-STD-1553B data bus has a practical data transfer limit of less than 1 Mbps. As store complexity increases, it is likely that the size of data files to be transferred will increase, with the associated increase in the time to transfer those files. In response to this, AS5653 (High-speed Network for MIL-STD-1760) has been developed which provides a Fibre Channel-based data network for transferring data over a copper media at a rate of 1.0625 Gigabaud.

As Fibre Channel is a commercial data transmission standard, several data protocols have been developed to satisfy the transfer of different data types. For example, ANSI INCITS 356-2002 Fibre Channel – Audio Video is used to transfer digitally encoded audio and video streams (note that the existing audio and video capabilities of MIL-STD-1760 are for analogue signals only). ANSI T11.3 Project 1648-DT Fibre Channel – Avionics Environment Protocol for MIL-STD-1553B Notice 2 also enables the transfer of MIL-STD-1553B formatted messages. This is particularly useful as the adoption of AS5653 provides a path to migrate the tried and trusted integrity of the MIL-STD-1760 protocols from the MUX A and MUX B interfaces to a very high-speed alternative.

6.4 Standardisation Conclusions

The examples described so far provide a basic physical/electrical infrastructure for weapons integration but do not address the detailed systems integration aspects, although, for example, MIL-STD-1760 does provision a set of standard data parameters which are used to simplify the number, type and format of data that stores can use. This means that once the aircraft system is able to generate the parameters, then they can be used to satisfy the data requirements of all compliant stores. However, even with these provisions, there may still be a significant effort required to develop system software changes required by the new store, not least because the standard does not dictate attributes such as update rates, allowable data latencies or how the parameters should be packed into data messages.

When integrating a new store, various system components will need to be modified to introduce control functionality. Also, although MIL-STD-1760 mandates certain standard data messages, it also permits systems integrators the freedom to use non-standard entities. This effectively works against re-use of system interfaces. MIL-STD-1760 does not define in which words specific data will be transmitted. For example, for a particular weapon, target position data may be located in data message number 14. For a different weapon, these data may be located in message 16 with message 14 being used for defining a route waypoint. Clearly, this would increase the complexity of operating the two weapons simultaneously.

In an attempt to reduce this problem, standardised data formats in addition to the provisions of MIL-STD-1760 have been introduced (particularly in the United States) with the aim of the aircraft provisioning a superset of message transactions, with the contents of each message being fixed. The store is then only allowed to use the subset of message transaction required to implement its pre-launch control. This will be explained further in Chapter 13.

Whilst adopting the ethos of standardisation can drive store designs away from being optimal, the compliant implementation of standardised interfaces brings other advantages such as greater interoperability across platforms.

Further Reading

Society of Automotive Engineers. (2008) High-Speed Network for MIL-STD-1760. *Aerospace Standard 5653*, Society of Automotive Engineers, Warrendale.

United States Department of Defense. (1978) Digital Time Division Command/Response Multiplex Data Bus. *Military Standard 1553B*, United States Department of Defense, Philadelphia.

United States Department of Defense. (2004) Aircraft/Store Electrical Interconnection System. *Military Handbook MIL-HDBK-1760A*, United States Department of Defense, Philadelphia.

United States Department of Defense. (2007) Aircraft/Store Electrical Interconnection System. *Military Standard 1760E*, United States Department of Defense, Philadelphia.

7

Other Weapons
Integration Standards

7.1 Chapter Summary

Having introduced standardisation in the previous chapter and reviewed MIL-STD-1760, a number of other important weapon interfacing standards will be described such as the Miniature Mission Store Interface (MMSI) standard (AS5725) and the standard for the Interface for Micro Munitions (IMM) (AS5726).

Over the past ten years, the weapons integration community has also started to develop a number of standards aimed at easing the systems integration activities and fostering greater interoperability. This chapter will also describe these standardisation documents which include the Generic Aircraft–Store Interface Framework (GASIF), the Mission Data Exchange Format (MiDEF) and the Common Launch Acceptability Region Approach (CLARA).

7.2 AS5725 Miniature Mission Store Interface

7.2.1 Interface Points

As the MMSI is designed for a smaller class of stores, there is a need for the interface to be compatible with a number of interfacing options. AS5725 includes two specific interface arrangements. The first, shown in Figure 7.1, is a multi-store carriage system which allows connection to individual Miniature Mission Stores either by blind-mate connectors or via an umbilical cable. It is assumed in the figure that the Miniature Store Carriage System (MSCS) will be carried on a standard aircraft station and be connected via a MIL-STD-1760 interface. However, should a MSCS be part of a smaller platform, it is acceptable for the MIL-STD-1760 interface not to be used and the interfaces for the miniature munitions provided directly.

Aircraft Systems Integration of Air-Launched Weapons, First Edition. Keith A. Rigby.
© 2013 John Wiley & Sons, Ltd. Published 2013 by John Wiley & Sons, Ltd.

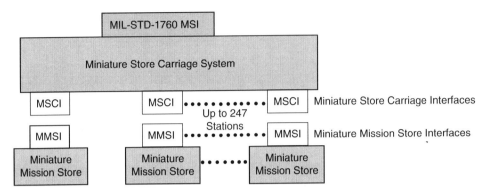

Figure 7.1   Overall Miniature Mission Store Interface arrangement

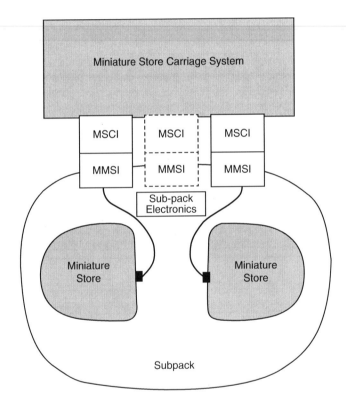

Figure 7.2   Sub-pack carriage arrangement

The MSCS provides the Mission Store Carriage Interface (MSCI), which connects to the MMSI on the store.

Some physical configurations of Miniature Mission Stores may require them to be packaged together in order to assist in their safe release from the launch aircraft. AS5725 therefore allows the alternative carriage arrangement of Miniature Mission Stores housed inside a sub-pack as shown in Figure 7.2.

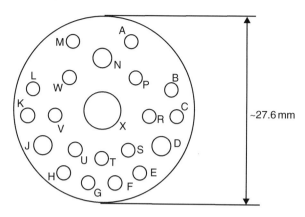

Figure 7.3 AS5725 connector insert arrangement

In this arrangement, the MMSI for each Miniature Mission Store is located on the sub-pack, and an additional MMSI is provisioned for the control of the sub-pack electronics. Each Miniature Mission Store is connected to its associated MMSI via an umbilical cable, the requirements of which are not defined by the standard (i.e. this is left to the WDO to specify).

7.2.2 Connector

AS5725 defines a single signal set allocated to a single connector. As Miniature Mission Stores are expected to be of a lower mass than traditional stores, the high break-force required to separate the MIL-STD-1760 connector during store release was deemed to be excessive. As the lanyard release mechanism force and the friction caused by the large number of contacts in the MIL-STD-1760 connector are significant, there was a need to use a different connector which has a significantly lower break-force. Part of the solution was to adopt a signal set containing fewer signals than the MIL-STD-1760 interface, thus allowing the use of a physically smaller connector. Figure 7.3 shows the connector insert format adopted by the standard.

7.2.3 Signal Set

The experience gained with the MIL-STD-1760 interface led the group defining AS5725 to reduce the number of options provided by the standard and to define a reduced signal set. In addition to this, four interface classes are defined. These are Class I (the full interface set without the Auxiliary Power interface), Class IA (the full interface set including the Auxiliary Power interface), Class II (the full interface set without the High Bandwidth, Fibre Channel and Auxiliary Power interfaces) and Class IIA (the Class II signal set plus the Auxiliary Power interface). The full signal set is outlined in Table 7.1.

An additional feature of the standard is that by the use of a simple adapter, the MMSI signal set can be used to control stores designed around the Joint Miniature Munitions Interface (an interface used by some early US miniature munitions).

Table 7.1 Miniature Mission Store Interface signal set

Miniature Munitions interface signal set		
Signal	Characteristics	Notes
Operating Power	28 V dc, 175 W power supply. Current/time profile defined by the standard	Store 'logic' supply for powering non-safety critical functions. Must be independently controllable from other supplies
Operating Power Return	Return line for Operating Power	Separate returns required for all power supplies
Safety Enable Power	28 V, 12.5 A dc power supply. Current/time profile defined by the standard	Store supply for powering safety critical functions. Must be independently controllable from other supplies
Safety Enable Power Return	Return line for Safety Enable Power	Separate returns required for all power supplies
Auxiliary Power	270 V, 1.5 A dc power supply. Current/time profile defined by the standard	Power source for stores that can make use of a higher supply voltage or where power needs are in excess of that available from the Operating Power supply. Used for store functions that are not safety critical
Auxiliary Power Return	Return line for Auxiliary Power	Separate returns required for all power supplies
Structure Ground	0.2 V maximum voltage drop when conducting an 8 A current. Must be capable of withstanding the over-currents defined for the interface power supplies	Electrical connection provided to minimise electric shock hazard for personnel. Must not be used as either a signal return or power return except under fault conditions within the store
Digital Data 1	10 Mbps digital time division command/response multiplex data bus. RS-485 transceivers employing the MIL-STD-1553B protocols	Point-to-point network configuration defined by AS5652. The adoption of MIL-STD-1553B protocols enables the higher level MIL-STD-1760 protocols to be employed in the control of Miniature Mission Stores
Down Fibre Channel	Fibre Channel copper interface with a data transmission rate of 1.0625 Gbaud. Down Fibre Channel is the aircraft to store path	Used as a high-data rate communications path for the transfer of digital video, digital audio or data messages replicating the MIL-STD-1760 and/or MIL-STD-1553B data bus protocols.
Up Fibre Channel	Fibre Channel copper interface with a data transmission rate of 1.0625 Gbaud. Up Fibre Channel is the store to aircraft path	See 'Down Fibre Channel'

(*continued*)

Table 7.1 (*continued*)

Miniature Munitions interface signal set		
Signal	Characteristics	Notes
High Bandwidth	Type A: 20 Hz – 20 MHz Type B: 20 MHz – 1.6 GHz 50 Ω impedance	Primarily used for the transfer of timing pulses (Type A signals) or GPS RF signals (Type B)
Safety Enable Discrete	Nominal 28 V dc low current (100 mA max) discrete signal. Return is via the Safety Enable Power return line	Provided to satisfy a safety function with consent only being enabled by the carriage system when the aircraft determines that safety criteria have been met
Platform Class	Three state discrete: • Open circuit indicates an airborne high-speed (>250 knots) platform • Nominal 28 V dc low current supply (referenced to Operating Power Return) indicates airborne low-speed platform (≤250 knots) • Short circuit to Operating Power Return indicates surface platforms	Used by the weapon to monitor the type of platform that is carrying the Miniature Mission Store. The carriage system conditions the signal depending on the type of platform
Store Mated	Nominal 28 V dc low current (100 mA max) discrete signal sourced by the carriage system	Used by the carriage system to monitor the electrically mated status of the associated store. Store connects this discrete to Operating Power Return
Carriage Mated	Nominal 28 V dc low current (100 mA max) discrete signal sourced by the Miniature Mission Store	Used by the store to monitor its electrically mated status with the carriage system. Carriage system connects this discrete to Operating Power Return

7.3 AS5726 Interface for Micro Munitions

7.3.1 Interface Points

By their very nature, micro munitions require the use of very small physical and functional interfaces on the munition. AS5726 does not define these interfaces, leaving it to the micro munition design documents and ICD to do so. AS5726 introduces the concept of separate interfaces for initial connection to the host and for separation during the launch actions. Figure 7.4 shows one example of how this may be realised in practice. The connection on the host is the Micro Munition Host Interface (MMHI), with the munition connector identified as the IMM. The connector at the launch separation point is specific to the micro munition design and is therefore outside the scope of AS5726.

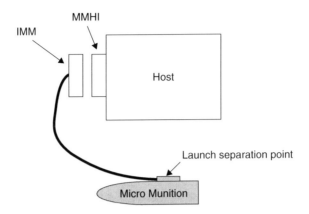

Figure 7.4   Example of IMM interfacing

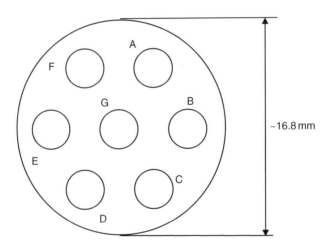

Figure 7.5   AS5726 connector insert arrangement

7.3.2   *Connectors*

The reduced signal set defined by AS5726 allows a single small connector to be employed.
Figure 7.5 shows the connector insert format adopted by the standard.

7.3.3   *Signal Set*

Two classes of interface (Class I and Class II) are provisioned by the standard. The difference
between the classes is the voltage level and maximum current which is available. The Micro
Munitions signal set is outlined in Table 7.2. A key point is that data and power share the same
physical interface connections.

Table 7.2   Micro Munitions signal set

Micro Munitions interface signal set		
Signal	Characteristics	Notes
Operating Power/Up Fibre Channel High	Operating Power for Class I interface is 28 V dc at 85 W Operating Power for Class II interface is 56 V dc at 200 Ω. Fibre Channel interface with a data transmission rate of 1.0625 Gbaud. Up Fibre Channel is the Micro Munition to Host path	Interface for transfer of power and high-data rate communications for the transfer of digital video, digital audio or data messages replicating the MIL-STD-1760 and/or MIL-STD-1553B data bus protocols
Operating Power Return/Up Fibre Channel Low	Return line for Operating Power / Fibre Channel interface with a data transmission rate of 1.0625 Gbaud. Up Fibre Channel is the Micro Munition to Host path	Interface for power return and high-data rate communications for the transfer of digital video, digital audio or data messages replicating the MIL-STD-1760 and/or MIL-STD-1553B data bus protocols
Safety Enable Power/Down Fibre Channel High	Safety Enable Power is 28 V dc at 85 W Fibre Channel interface with a data transmission rate of 1.0625 Gbaud. Down Fibre Channel is the Host to Micro Munition path.	See 'Operating Power/Up Fibre Channel'
Safety Enable Power Return/Down Fibre Channel Low	Return line for Safety Enable Power. Fibre Channel interface with a data transmission rate of 1.0625 Gbaud. Down Fibre Channel is the Host to Micro Munition path	See 'Operating Power Return/Up Fibre Channel'
Mated Status	Host connects this discrete to Operating Power Return via a 10 kΩ resistor. Mated status is detected by the host if voltage at this interface is greater than 20% of the Operating Power voltage and by the micro munition by similar means, using its internal power supply as the reference	Used by the Host and the Micro Munition to monitor the electrically mated status of the interface
Safety Enable Discrete	Nominal 28 V dc low current (100 mA max) discrete signal. Return is via the Safety Enable Power return line	Provided to satisfy a safety function with consent only being enabled by the Host when criteria for the employment of the Micro Munition have been met
Structure Ground	0.2 V maximum voltage drop when conducting a 3.8 A current. Must be capable of withstanding the over-currents defined for the primary interface power supplies	Electrical connection provided to minimise electric shock hazard for personnel. Must not be used as either a signal return or power return except under fault conditions within the store

7.4   Other Weapons Integration Standards

Whilst the standards discussed in the first part of this chapter and in Chapter 6 foster a level of interoperability between the aircraft and the store, they do not address the implications for the launch aircraft systems and software. Whilst it is unlikely that systems across every aircraft will ever employ the same electronic and software architectures, there is still a need to address aspects where a level of standardisation can be introduced. Three particular areas have been considered, aimed at bringing a level of interoperability to the systems integration aspects. These are the GASIF, the MiDEF and the CLARA. Each of these will be discussed in the following sections.

7.4.1   Generic Aircraft–Store Interface Framework

A significant consideration for an integration programme is the data interface between the aircraft and the store and how this is implemented in the computing architecture of the platform, store and intelligent carriage system. Recognising that there would be a need in the future to extend the capabilities of the data transfer medium in the MIL-STD-1760 interface, the SAE initiated the development of an Aerospace Information Report (AIR 5532, the GASIF) aimed at defining a common language which could be used when defining future data transfer standards. GASIF has been defined to comply with the Open Systems Interconnect Basic Reference Model defined by ISO/IEC 7498-1 (Information Technology – Open Systems Interconnection (OSI) – Basic Reference Model: The Basic Model).

The OSI Basic Reference Model contains seven layers, each of which provides a data communications service to the layer immediately above it. The top layer (the Application Layer) corresponds to the data used by the computing system, with lower layers progressively adding protocol information until at the lowest layer (the Physical Layer), is the interconnection medium (i.e. the hardware transceivers and cabling/optical fibres).

Figure 7.6 shows a simple example of peer-to-peer interfaces for an aircraft to Mission Store interface with an interconnecting Carriage Store.

The figure shows all seven layers of the OSI Reference Model in both the aircraft and the Mission Store, the interconnecting lines showing the logical, peer-to-peer interfaces. In this example, the Carriage Store provides a routing mechanism between the aircraft and the Mission Store, but only replicating the three lower layers of the model. Table 7.3 defines the purpose of the individual layers and identifies examples of the breakdown of the MIL-STD-1760 protocols relating to each layer. It should be noted that as the MIL-STD-1760 pre-dates the OSI Basic Reference Model, the protocols were not defined with the seven layer model in mind, and therefore some elements cannot be accurately mapped to some layers.

From Table 7.3, it can be seen that the MIL-STD-1553B data bus interface exists in the lower three layers. In the example in Figure 7.6, for the aircraft to maintain control of Mission Stores, the Carriage Store is required only to route MIL-STD-1553B messages; that is, from a data communications viewpoint, it only operates in the MIL-STD-1553B domain.

It has been noted that both the Miniature Munitions and High-speed 1760 interfaces employ the MIL-STD-1760 protocols but using a different physical media. With reference to GASIF, it can be seen that the introduction of a 10 MB/s digital time division command/response multiplex data bus (sometimes referred to Extended Bit Rate or EBR-1553 and defined by AS5652) as the Miniature Munition data interface only affects the Physical Layer of the

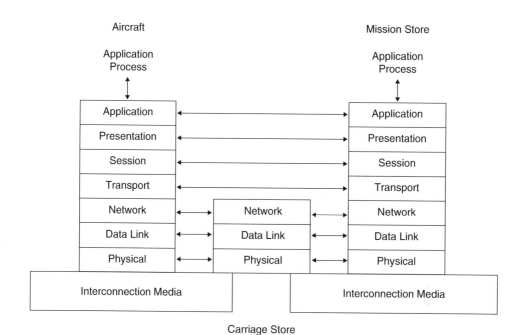

Figure 7.6 OSI Reference Model Mapping to MIL-STD-1760 protocols (aircraft interconnection with a Carriage Store and Mission Store)

Table 7.3 OSI Reference Model Mapping to MIL-STD-1760 Protocols

Layer	Definition
Application	The highest layer in the model which provides the means for the computer programme implementing the system functionality (Application Process) to access the data communications system. MIL-STD-1760 does not have any specific protocol elements which map to this layer although the Application Process is where the standard data entities exist, for example True Airspeed data and the Critical Control and Critical Authority words
Presentation	Defines the abstract format of data to be transferred between computer networks (e.g. the format of real data to be transferred from an aircraft to a store defined in MIL-STD-1760 by the linear data entities, for example Velocity(M) and Velocity(L) words)
Session	Responsible for establishing connections, synchronising data transfers and managing exceptions. The MIL-STD-1760 Mass Data Transfer protocol implemented using the Transfer Control and Transfer Monitor messages is a Session Layer protocol.
Transport	Data formatting and construction into packets. The MIL-STD-1760 Mass Data Transfer messages used to transport records and blocks exist within the transport layer
Network	MIL-STD-1760 defers to the MIL-STD-1553B Network protocols such as the rules for the use of mode codes
Data Link	For MIL-STD-1760, this is captured by the MIL-STD-1553B basic protocols, normally embedded in the protocol chip set
Physical	Data bus voltage levels, impedance and Manchester bi-phase waveforms employed by MIL-STD-1553B

model. However, the introduction a High-speed 1760 Fibre Channel interface affects the Physical layer, and, depending on which Fibre Channel protocol is adopted (e.g. FC-AE-1553, FC-AV, etc.), some of the higher layers will also be affected. GASIF therefore provides an effective means to understand the impact of such changes to the data transfer medium.

7.4.2   Mission Data Exchange Format

MIL-STD-3014 (Department of Defense Interface Standard for Mission Data Exchange Format – MiDEF) has been introduced to provide a standard means of formatting weapons-related mission data files. Prior to this standard, such files would be developed on a Mission Planning System for both the launch aircraft and the weapon and uploaded prior to the sortie using, for example, a data transfer cartridge. If a sortie required more than one target to be prosecuted or if there was a need to provide alternative targets, then several mission plans would need to be loaded to the aircraft prior to flight. If the launch aircraft needed to be re-tasked whilst in flight, then a new mission plan would need to be transferred either by voice communication, requiring the aircrew to manually insert new data into the aircraft's systems, or over a data link such as Link-16. Should the weapon load-out consist of more than one type of smart weapon, then the planning would have to generate potentially very different data files. Mission data for either the aircraft or the weapon would not normally be standardised, with each defining its own formats and message content. This adds complexity and can lead to an inefficient use of valuable data bandwidth on communication channels. With a greater emphasis on the prosecution of time-sensitive targets, the need for a reduction in the time required for transferring new mission plans and the added flexibility needed requires a level of standardisation across Mission Planning Systems, aircraft and weapons.

MIL-STD-3014 is a Transport Layer protocol aimed at establishing a common mission data file format for the end-to-end weapon system which enables mission plans to be created, modified, transferred and interpreted using common data items.

Whereby a mature standard such as MIL-STD-1760 includes all available data items within the standard document, MIL-STD-3014 refers to an Internet-based registry. The registry was originally populated with the relevant data entities from MIL-STD-1760 and then supple-mented by a number of other 'user-defined' entities from specific programmes. Where a new programme requires a data item which is not currently defined in the registry, an application can be made to the registry administrators (currently the US Navy) for the new entities to be added to the registry. This enables programmes to quickly get the approval for new mission data items without waiting for the 'paper' standard to be revised.

Figure 7.7 shows an example of the logical organisation of a typical mission data file.

It can be seen that a typical file can be made up of a number of elements defined by the standard. These are Module, Header, a Primitive element and a Concatenated element. A MiDEF file consists of one or more Modules. A typical Module consists of a Header followed by as many elements as required to define the mission data for the aircraft and its weapons. Modules can contain other modules if required.

The basic element of the file is the Primitive Data Element, which is a single data element from the Internet-based registry, for example, the MIL-STD-1760 data entity Primitive 'XFER ALN PLTFM VEL NRTH'. This is defined in the registry as the Aircraft velocity north most significant and least significant parts which are defined by the Velocity(M) and Velocity(L) data entities.

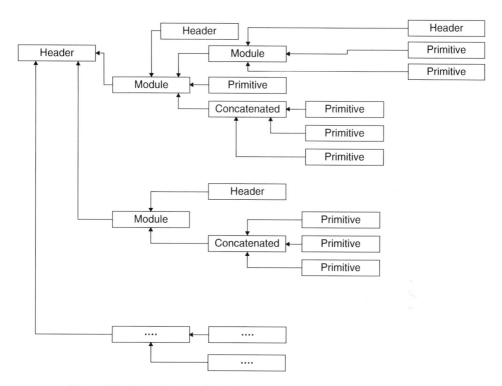

Figure 7.7 Example of the logical organisation of a typical mission data file

The registry also allows a number of Primitive elements to be grouped into a Concatenated Data Element. For example, the GENERAL VEL NED (velocity vector, north, east and down components) element consists of three Primitive GEN VEL-M (Velocity(M) data words) elements.

MiDEF provides a standardised protocol for mission data but with sufficient flexibility and extensibility such that the design of future weapons capability is not constrained.

7.4.3 Common Launch Acceptability Region Approach

Chapter 5 briefly introduced the concept of having a single LAR equation that could be modified for different weapons purely by changing the coefficients in the equation. Such an approach has been developed by the SAE as recorded in a number of publications (notably AIR5682 (CLARA ICD), AIR5712 (CLARA Rationale Document) and AIR5788 (CLAR Truth Generator ICD for the CLARA).

CLARA encompasses the activities which the WDO will undertake during the development programme to define the actual dynamic performance of the weapon and to generate the coefficients which can then be used by the ADO in configuring a common algorithm in the aircraft's Weapon Aiming Computer. The approach is founded on standardising the way in which the algorithm coefficients are developed by the WDO. Within the weapon's development environment, the LAR coefficients are calculated by running a high fidelity model of the weapon's flight dynamics (the Weapon Truth Model). This data set can then be used to translate aircraft flight parameters into a prediction of the ability of the weapon to hit its target. CLARA therefore also standardises the format of input data to the Weapon Truth Model and defines the required output data and the units and data format.

Using standard data inputs and outputs enables a common LAR algorithm to be defined, albeit this will be an approximation of the Weapon Truth Model for the actual weapon. However, as new weapon programmes adopt CLARA, the only difference in the algorithm will be the coefficients generated by the CLARA process. This enables a common LAR algorithm to be developed and used by the Mission Planning System, the launch aircraft and the training systems.

Further Reading

International Standards Organisation. (1994) Information Technology – Open Systems Interconnection – Basic Reference Model: The Basic Model. *ISO/IEC 7498-1*, International Organization for Standardization, Geneva.

Society of Automotive Engineers. (2003) Generic Aircraft-Store Interface Framework. *Aerospace Information Report AIR 5532*, Society of Automotive Engineers, Warrendale.

Society of Automotive Engineers. (2005) 10 Megabit/sec Network Configuration Digital Time Division Command/ Response Multiplex Data Bus. *Aerospace Standard 5652*, Society of Automotive Engineers, Warrendale.

Society of Automotive Engineers. (2005) Common Launch Acceptability Region Truth Data Generator Interface Control Document for the CLAR Approach. *Aerospace Information Report AIR 5788*, Society of Automotive Engineers, Warrendale.

Society of Automotive Engineers. (2007) Common Launch Acceptability Region Approach Interface Control Document. *Aerospace Information Report AIR 5682*, Society of Automotive Engineers, Warrendale.

Society of Automotive Engineers. (2008) Interface Standard, Miniature Mission Store Interface. *Aerospace Standard 5725*, Society of Automotive Engineers, Warrendale.

Society of Automotive Engineers. (2008) High Speed Network for MIL-STD-1760. *Aerospace Standard 5653*, Society of Automotive Engineers, Warrendale.

Society of Automotive Engineers. (2009) Interface Standard, Interface for Micro Munitions. *Aerospace Standard 5726*, Society of Automotive Engineers, Warrendale.

United States Department of Defense. (2007) Mission Data Exchange Format. *Military Standard 3014*, US Department of Defense, Philadelphia.

8

Interface Management

8.1 Chapter Summary

Weapons integration programmes require the ADO and the WDO to collaborate. The interface between the aircraft and the weapon must therefore be agreed by the two parties and documented in an ICD. This chapter will give an overview of the type of information that has to be agreed in an ICD but will also detail the differing approaches between the United States and Europe in controlling the interface data and managing the process to agree the ICD. Where a weapon is to be integrated across a number of platforms, there may be a need for several programmes to develop and agree ICDs simultaneously. This provides a significant organisational challenge. Strategies such as the need for an Interface Control Plan (ICP) and multi-programme Interface Control Working Groups (ICWGs) will also be considered. Finally, this chapter will discuss an effective management process for controlling and reducing integration risk.

8.2 Introduction

Whenever a weapon has to be integrated with an aircraft, it is likely that at least two separate organisations will be involved: the ADO and the WDO. This means that a robust interface management process will be required between the two organisations to ensure that the integration programme runs smoothly. An important consideration is the control of the different interfaces between the platform and the store which must be documented in an ICD and agreed between the ADO and WDO.

The differing interface management approaches between the United States and Europe will be explained and a robust process for managing integration risk will be explored. The chapter

Aircraft Systems Integration of Air-Launched Weapons, First Edition. Keith A. Rigby.
© 2013 John Wiley & Sons, Ltd. Published 2013 by John Wiley & Sons, Ltd.

will also give an overview of the type of information contained in the ICD and will discuss AS5609 which defines a common ICD format aimed at fostering greater commonality in the type and level of detail of information that needs to be agreed.

8.3   Management of the Aircraft/Store Interface

The complex nature of the integration programme means that a structured management organisation is required. Whilst the methods for managing integration programmes differ between Europe and the United States, the management principles will be based on some form of ICWG).

For a simple integration programme, a single ICWG will handle all contractual, financial and aircraft/store technical interface details. However, for more complex programmes, there may be a need to divide responsibilities for managing the interface across two or more management levels with the main ICWG taking responsibility for the management and control of all technical interface details, ensuring that all relevant parties reach agreement on the content of the ICD and its related documents. Generally, the ICWG participants are the ADO and WDO, with representatives from the Contracting Agency.

During the integration programme, there may be occasions where the ADO and WDO cannot reach agreement. For this, there needs to be a higher-level executive which can arbitrate. In reaching a decision, the executive will need to consider all aspects of the problem. This will include an assessment of the technical merits of differing solutions proposed by the DOs, but must also consider wider programme issues such as the cost and timescale implications of changes to both the aircraft and the store needed to accommodate a workable solution. The executive would be governed by the Contracting Agency.

Where a store is to be integrated onto more than one aircraft, there are two approaches which are generally adopted. These are broadly the European and the American approaches.

In Europe, a second ICWG will be initiated to control the second aircraft's interface with the store. This is necessary to maintain the confidentiality of the national technical capability of each aircraft from the other ADO. As the WDO is usually the only common party between the integration programmes, it is the WDO which assumes responsibility for ensuring that the interfaces agreed for the different aircraft can all be satisfied by the single store design. In practice, this means that there will be separate ICDs for each aircraft which integrates the store.

As the ICWG is fundamentally a technical and programme management organisation, there will also be a need for specialist subsidiary groups to be instigated. Such technical subgroups could cover specialist areas such as aerodynamics, or electromagnetic compatibility, and will meet as required to develop sections of the ICD or to debate integration problems relating to their specialist area.

Where there are different nations involved in the integration of a single weapon across several platforms, then the executive body overseeing the ICWGs will contain representatives from all ADOs, the WDO and the national Contracting Agencies.

Figure 8.1 shows an example of a European ICWG organisation.

In the United States, the aircraft and weapons are usually designed and developed by US companies. This means that there is little concern over protecting national sensitivities. As a

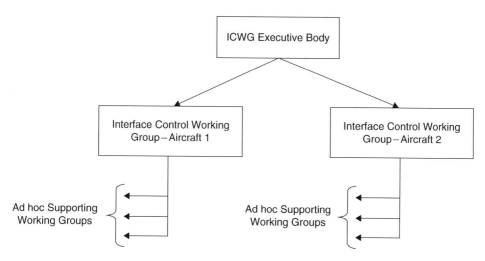

Figure 8.1   Example of the ICWG Organisation

result, it is possible to combine the ICWGs concerned with the integration of the store onto several platforms into a joint body, the Joint Interface Control Working Group (JICWG).

The JICWG consists of a number of roles, these being the Chairperson, the Secretariat (which undertakes the daily running of the JICWG), Interface Managers/ Participants (who are the representatives from each of the programmes) and Advisors (specialists who offer advice to the JICWG). The JICWG has representatives from the Contracting Agency, the WDO and the ADOs for each platform which is to integrate the store. During the store's System Development and Demonstration phase, the JICWG chair and secretariat are provided by the Contracting Agency. However, once the store programme transitions into the production phase, the WDO would normally assume responsibility for the secretariat role.

Where changes to the interface are proposed, these are reviewed and agreed by all parties to the JICWG. However, should the JICWG be unable to reach agreement, the issues are passed to the first higher-level executive, the Interface Management Board (IMB). The IMB consists of senior representatives from the industrial organisations involved in the JICWG and also members of the Contracting Agency. Should the IMB be unable to resolve the issue, then the final referral is to the Interface Control Steering Group, a body consisting of senior representatives from the Contracting Agency's hierarchy. An example of the organisation is shown in Figure 8.2.

The complexities of having to get multiple stakeholders to agree the interface require careful management. It is for this reason that in the United States, the overall process is governed by an ICP. The ICP defines the roles, responsibilities, communication principles and documentation applicable to each of the organisations involved in the programme. It also defines the roles and responsibilities of the interface control organisations.

Within the remainder of this chapter, ICWG will be used to refer to the body responsible for the control of the interface, accepting that the principles discussed later would be implemented using one of the approaches just described.

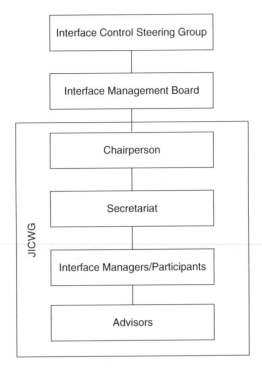

Figure 8.2   Example of the JICWG Organisation

8.4   Approaches to Interface Documentation

The approach to documenting the aircraft/store interface also differs between the United States and European programmes. In the United States, the ICD defines the interface for a specific weapon. The document establishes, defines and controls the functional, logical, physical/mechanical, electrical and environmental aspects of the interface between the weapon and an aircraft. Being written from the weapon's viewpoint, the ICD documents the ideal interface which must be provided if the weapon is to meet its required performance when integrated with an aircraft. The ICD therefore comprises a core document which is supplemented by a group of annexes covering each of the aircraft with which the weapon is integrated.

Aircraft annexes typically have a unique annex designator which identifies the aircraft type and variant (e.g. F-15E) and may provide a country indicator for platforms which have been exported.

Each annex documents the aircraft's side of the interface and follows the format of the core document but records the aircraft's actual implementation of the interface. The annexe, ideally, would comply with all the mandated interface requirements in the core document. However, should the aircraft be unable, for whatever reason, to provide the desired interface, then this must also be recorded in the annexe such that the implications on weapon performance can be determined and agreed with the Contracting Agency. The annex would also provide timelines which show the actual key timing relationships associated with the interface.

In the United States, there are a relatively large number of aircraft and weapon types creating greater opportunities for new integration programmes. In Europe, where there has been a significant rationalisation of the defence industry such that there are relatively few indigenous aircraft and weapons, there are few integration programmes. This rationalisation has led to the emergence of a relatively common approach to the generation and management of ICDs.

In the majority of European integration programmes, the ICD is written and managed by the ADO with a significant contribution from the WDO. The ICD defines the interface between a single aircraft type and a single store type, for example, the integration of the Paveway™ IV guided bomb with the Tornado GR4 aircraft. The interface between Paveway™ IV and another aircraft such as Typhoon would be defined in a completely separate ICD.

As a result of the multi-national nature of European industry, several different ICD formats have emerged, each with its own merits. Variations range from a single volume containing definitions for all the interface characteristics, through to a two-volume ICD, where the first volume records aspects of the interface which are less likely to change (e.g. the aeromechanical or physical details), with a second volume dealing with those areas which are the subject of frequent changes during the integration programme (e.g. avionic and operational interface details).

Either of the approaches are valid methods of formatting an ICD. In reality, the actual approach adopted will be dictated by national preferences.

8.5   Interfaces Documented in the ICD

The role of the ICD is to contain all of the detail necessary to completely define the interface between an aircraft and the store. This may be by the inclusion of interface details within the ICD document or by reference to other documents. Where other documents are referenced, then these become a part of the interface definition, for example, a reference to the MIL-STD-1553B electrical signal characteristics. However, where reference is made to another document, then it is important that the exact version or revision number of the document is unambiguously referenced to avoid the interface definition being compromised should the custodian of the referenced document change the content, thereby unwittingly invalidating the ICD.

The ICD identifies what each side of the interface must provide. It is therefore incumbent on the DOs to ensure that their side of the interface is correctly implemented. The ICD must define the actual interface rather than what one or other of the DOs would prefer it to be. That is, the ICD contains statements of fact regarding the aircraft and the store.

Statements in the ICD are usually expressed as 'shall' statements so are generally accepted as being requirements. Indeed, it will be necessary for the DOs to validate their side of the interface and provide documentary evidence to verify that it has been correctly implemented. However, in order to manage integration risks, the ICD 'requirements' should be constrained such that no new requirements are defined which could change the previous designs of the aircraft or store unless there is no alternative. Any such changes must be agreed within the ICWG such that there is an awareness of the consequential effects of decisions taken when compiling the ICD. In order to avoid such instances arising, the ICD may be included by the Contracting Agency in the contract documentation placed on both the ADO and the WDO. In some cases, the Contracting Agency may also be a signatory to the ICD, therefore requiring the Agency to be involved in the decision-making process for all changes, regardless of magnitude. The ICD, as a minimum, carries the authorisation signatures of both the ADO and WDO.

As noted earlier, the essential interface parameters will be documented using 'shall' statements. Other terms will also be used such as 'should', 'must', 'will' and 'may'. For the specific integration programme, the meaning of each term is recorded in the ICD. Table 8.1

Table 8.1 Example of the meaning of ICD terminology

Term	ICD definition
SHALL	The word SHALL in the ICD text expresses a requirement. Departure from such a requirement is not permissible without formal agreement between the ADO and the WDO
SHOULD	The word SHOULD in the ICD text expresses a system objective. Failure to meet a 'Should' statement is to be justified, and this needs to be accepted by the ADO and the WDO
MUST	The word MUST in the ICD text is used for legislative or regulatory requirements (e.g. Health and Safety) with which both the ADO and the WDO are mandated to comply. It is not used to express a requirement of the ICD
WILL	The word WILL in the ICD text expresses a provision or service or an intention in connection with a requirement of the ICD. Both parties can reasonably expect to rely on such service or intention
MAY	The word MAY in the ICD text expresses a permissible practice or action. It does not express a requirement of the ICD

provides a typical meaning for these terms, although it should be noted that the exact meanings will be agreed within the ICWG.

The ICD does not exist at a single level in the V-Diagram (see Figure 3.5) as it is a living document that will be subject to continuous change during the integration programme. This means that formal configuration control and change control processes must be applied such that the ICD cannot be changed without the agreement of the ICWG (as a minimum, both DOs). The recognised method of controlling change is to use an Interface Change Note (ICN – although it could have a different title on different programmes or in different countries). The actual process employed would normally be defined in the ICD.

The ICN is compiled by the party wishing to change the ICD and distributed for agreement by the ICWG. Clearly, the ICN as initially defined may not be acceptable to all parties, and therefore there will need to be a negotiation to reach an acceptable compromise. Only when the full process has been completed and ICWG agreement reached will the ICD be updated. Certainly in the early stages of an integration programme, when it is likely that many changes will be introduced, it would be normal practice to incorporate several ICNs in a single revision of the ICD.

In some cases where the change to the interface has an impact on the cost or schedule of an integration programme, there will be a need to involve the Contracting Agency in the decision process. On occasions when the ICWG finds it impossible, for whatever reason, to agree interface technical details, meetings between the relevant stakeholders (as a minimum, the Contracting Agency, the ADO and the WDO) will be convened to assist in reaching mutual agreement. It is for this reason that a formal interface management process must be defined and agreed at the outset of the integration programme.

As noted earlier, between an aircraft and a store, there are many potential interfaces which need to be agreed by the ICWG. AS5609 has been introduced to standardise the format and level of content that should be captured by the ICD.

For a simple interface, a single volume ICD may be sufficient to record all interface parameters. But as noted earlier, for complex interfaces, where there is likely to be significant change introduced during the integration programme, a two-volume version is recommended. AS5609, which is based on best practice of ICD formats from both Europe and the United States, reflects this approach placing data for the parts of the interface which are less likely to change in the first volume of the ICD, with more volatile parts in a second volume. Table 8.2

Table 8.2 Data captured in the ICD

Section number	Contents	Overview
Volume 1		
1	Introduction	This section forms the introduction to Volume 1 of the ICD and provides an overview of the content of the volume. Section 1 identifies the process for controlling and updating this volume of the ICD. Abbreviations and terms (such as 'shall', 'may', etc.) are included. The purpose of the store, the aircraft or the combination is also defined in this section, and for a store-based ICD, the relevance of the document to compatible aircraft is also recorded
2	Applicable Documents	This section lists all documents that are fundamental to the definition of the interface data in Volume 1 and defines the order of precedence between those documents and the ICD. This section includes the relevant issue or revision of the documents used during compilation of the ICD to prevent the interface data being invalidated should a referenced document be updated by its custodian
3	Mechanical Interface	This section defines the characteristics of the physical attributes of the interface between the aircraft and the store such as attachment locations, alignment information and lanyard attachments. Other physical attributes such as moments of inertia, centre of gravity, loads data, stiffness data are also recorded. This section also identifies any considerations for store loading such as access to physical connections and clearance for the use of specified loading equipment. The units of measurement, scaling and scaling factors to facilitate interoperability are also included
4	Electrical Interface	This section defines all relevant electrical, electromagnetic and optical signals that are a part of the interface. Wherever possible, reference is made to existing standards and definitions in order to avoid multiple specifications. The connector type, layout and contact assignment are defined followed by a general description of all signals which are likely to include power, analogue signals, digital signals and so on
5	Aerodynamic Interface	This section defines all aerodynamic parameters (e.g. drag, altitude limits, incidence limits, carriage envelopes, release envelopes, jettison envelopes, etc.) relating to store carriage, release/jettison and post-launch manoeuvres in the vicinity of the aircraft
6	System Safety	This section defines all safety issues for ground operations, captive carriage, release/jettison and post-launch. As safety issues can differ depending on the phases of the mission, this section of the ICD identifies relevant safety considerations for each phase (e.g. for ground operations, issues such as chemical substances, electrical installations, electromagnetic radiation, etc. would be covered)
7	Environmental Conditions	This section defines relevant influences caused by environmental impact. This section will cover all aspects of the operational environment conditions such as rain, mechanical shock, acceleration, vibration, humidity and electromagnetic environmental effects

(continued)

Table 8.2 *(continued)*

Section number	Contents	Overview
8	Aircraft/Store Interface Specific	All interface issues that have not been captured elsewhere in the ICD (e.g. cooling, radar signatures, rocket plume characteristics, etc.) are defined by this section
9	Aircraft/Store Post-Launch (Data Link) Interface	This section defines the post-launch interface used to communicate with a weapon post-launch (e.g. guidance commands, sensor data and battle damage information). This section would include a definition of the electrical part of the post-launch communication. The logical definition is included in Volume 2 of the ICD
10	Support Interface	This section defines aspects relating to the testing and ground handling of the store. For example, this section could address issues relating to ground test equipment or store loading/unloading/arming considerations
Volume 2		
1	Introduction	This forms the introduction to Volume 2 of the ICD and provides an overview of the content of the volume. Section 1 identifies the process for controlling and updating this volume of the ICD. Abbreviations and terms (such as 'shall', 'may', etc.) are included
2	Applicable Documents	This section lists all documents that are fundamental to the definition of the interface data in Volume 2 and defines the order of precedence between those documents and the ICD. This section includes the relevant issue or revision of the documents used during compilation of the ICD to prevent the interface data being invalidated should a referenced document be updated by its custodian
3	Functional Interface	This section is the core of Volume 2 and details the operation of the store as a part of an integrated configuration of the aircraft and the store. The co-ordinate systems employed by the aircraft and store would be defined as would the timelines for data exchange and the activation/deactivation of interface signals. State transitions and the actions required to transition between system modes are also detailed in this section
4	Communications Interface	This section defines the detailed characteristics of the data parameters (bit meanings, word formats and sequences, etc.) passed across the interface and the implementation of protocols. Any deviations from the standard protocols or the implementation of optional features would be specified
5	Post-Launch Communications Interface	This section defines the protocols implemented on the post-launch communications interface along with any deviation from standard protocols or optional features which are implemented. The detailed characteristics of the data parameters (bit meanings, word formats and sequences, etc.) passed across the interface are also defined
6	Signal Format Sheets	This section contains data sheets which define the format of signals in the aircraft/store direct interface
7	Post-Launch Signal Format Sheets	This section contains data sheets which define the format of signals in the post-launch interface

shows the recommended format and identifies the type of data that would be captured in each section of the ICD.

Whilst Table 8.2 outlines the purpose of each section of the ICD, AS5609 provides detailed examples of the types of data which would normally be included in an ICD.

8.6   Controlling the Interface of Store Variants

When integrating a new store type, there is always a need to employ variants of the operational store. This could be because the store is still being developed and operational units are not available. However, it is more likely that the special requirements needed for handling and operating operational stores, particularly if they contain explosive devices, make it impractical to use these in areas such as the SIL. Some store variants such as dummy stores are often used by the operational user for training loading and arming crews. Other variants such as an environmental data gathering store are only used to support the integration programme. These two categories of stores are termed 'Service Variants' and 'Development Variants'.

There are two primary ways in which a programme can manage the definition of the interfaces of these variants. The first and most common approach is for the ICD to define the interfaces between the aircraft and the Service Variants only. The main body of the ICD defines the interface applicable to the operational store, with each of the Service Variants being covered in separate appendices. Each appendix follows the same paragraph numbering as the main body of the ICD, but contains definitions in only those paragraphs where the content differs.

Development Variants may have significantly different interfaces (particularly the electrical and system interfaces) from that of the operational store. In this case, the interface for each of the Development Variants is defined in separate, self-contained documents: Interface Agreement Documents (IADs). An IAD defines the complete interface between the aircraft and one of the variant types, for example an Environmental Data Gathering store. The IAD is identical in structure to the operational store's ICD and is controlled in the same way as the ICD. Each Development Variant will have its own IAD. As noted earlier, the ICD is often a document which is included in the formal integration programme contract. IADs are not usually part of the contract, and the interface is usually fixed and unlikely to change during the integration programme. Whilst there is still a need for both the ADO and WDO to agree the contents of the IAD, because there is not a specific link to the integration programme's contract, the means of control can be less formal.

The use of an ICD (for Service Variants) and IADs (for Development Variants) allows a more flexible approach to be used and is particularly advantageous where either the aircraft or the store is in its development phase.

An alternative approach to managing the interfaces of different variants is where all of the variants' interfaces are defined within the single ICD, with the interfaces for each of the Service and Development Variants being defined in separate appendices. This approach allows the interfaces for all variants of a single store type to be defined in a single document. This can be a useful approach for when the store and aircraft interfaces are mature or are very simple in nature.

8.7   Information Exchange between Design Organisations

An integration programme requires a great deal of information to be exchanged between the DOs. Some data, for example, strength data relating to an existing missile, may be available at the start of the integration programme. However, there will be much data which have to be generated as a part of the integration activities. A typical example could be information relating to the safety of the weapon when it is operated in the aircraft's electromagnetic environment. Recognising that not all information will be available at the beginning of an integration programme, there is a need to establish at the outset a programme of two-way information exchanges which details the specific data to be exchanged and when it is planned to be exchanged. This Data Requirements List (DRL) would normally be included in the integration contract.

It is usual for the data exchanged to be reviewed by the receiving DO. The data may unwittingly be incomplete, and this will lead to comments being raised and the delivered data being updated. This adds cost into the programme and so should be avoided wherever possible by carefully defining the expected content of each DRL item as a part of the integration contract.

8.8   Process for Managing Integration Risk

It is clear from the discussion so far that an aircraft/store integration programme is a complex interaction involving the ADO, the WDO and the Contracting Agency. Therefore, it is essential that for an integration programme to be successful, careful management of integration risk is required. For this, there needs to be a process which is adopted by all parties. Such a process is outlined in this section.

Traditionally, it has been the ICD which has been central to the overall route from agreeing the integration requirements, through implementation in both the aircraft and the store design, to testing and analysis and finally to certification of the aircraft/store combination. However, the ICD will probably be immature at the start of the integration programme, and with the complexity of integration programmes increasing, significant integration issues and disputes which serve to extend the integration programme and increase integration costs have become prevalent. Such problems can become a major barrier to efficient integration programmes. Therefore, programmes should not rely on the ICD as the sole vehicle for controlling the integration programme; a more robust stores integration process is therefore essential.

It should be remembered that requirements for the integration of the store with the aircraft flow out of Contracting Agency's capability needs and the needs of both the aircraft and the weapon. As noted in the earlier chapters on the systems engineering process, it is essential to address integration requirements early in the integration programme in order to avoid problems later, when the impact will be greatest.

Figure 8.3 identifies a process for managing interfaces based on the V-Diagram process in Figure 3.5 (the shaded boxes are those which are directly traceable to the V-Diagram).

Central to reducing integration risk is the development of the IRS. The IRS captures all requirements derived from the Contracting Agency's Statement of Requirements which relate to the integrated capability of the aircraft and the store. Effectively, the IRS requirements will cover the high-level aspects of the integration programme. As the IRS requirements translate

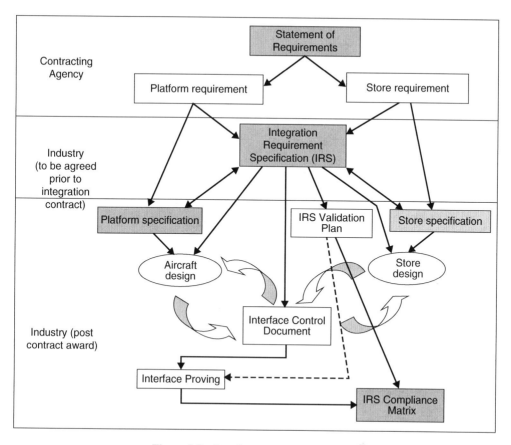

Figure 8.3 Interface management process

into the activities that will be undertaken by the DOs, the document must be agreed by both parties. As the IRS is derived from the Contracting Agency's requirements, there is also a need for the Contracting Agency to agree the IRS content.

In order to reduce integration risks, the IRS should be developed ahead of the acceptance of any development or integration contracts by the ADO and WDO. This enables only the essential requirements for both the aircraft and the store to be captured and the means for demonstrating compliance jointly developed and agreed early. The benefit of doing this is that risks can be identified pre-contract and mitigation of the consequences of the risks agreed between the two DOs and the Contracting Agency. The discipline of compiling the IRS will identify any requirements that either cannot be met or can only partially be met by one or both DOs. With the involvement of the Contracting Agency, the solution to requirements which cannot be agreed can be debated and an appropriate way forward determined which can be agreed by all parties before the Contracting Agency places development/ integration contracts on the DOs. This approach will enable effective risk-reducing trade and cost studies to be undertaken pre-contract and the associated aircraft and store specifications to be adapted as needed to reflect the agreed requirements ahead of the main integration contract.

Where an existing store is to be integrated, it is desirable to avoid any requirements which will cause the store design to change, unless the Contracting Agency is satisfied that a change is required to satisfy the capability needs. Agreeing requirements early ensures that no new requirements are generated unless absolutely necessary (e.g. the need to operate in a harsher environment). This will lead to better forecasting and mitigation of the consequences of risk that are often the cause of programme delays and cost escalation.

The IRS process also provides a mechanism for addressing the coherence of safety requirements by capturing a single set of certifiable requirements that are placed on both the aircraft and weapon systems.

Prior to the IRS, both the ADO and WDO will each have some understanding of what is required, including the Contracting Agency's requirements and Concept of Operation for the new aircraft/store combination. The store will have a certain level of definition whether it is a brand new design or off-the-shelf. The aircraft will have its existing capabilities and the environment it provides to the stores. In order to generate the IRS, it will be necessary to compare what the store requires with what the aircraft provides with the aim of identifying any inconsistencies. In effect, by documenting this comparison, areas that match between the aircraft and the store become the IRS requirements. Where there are incompatibilities, wherever possible these need to be reconciled. Where requirements cannot be reconciled, then a mitigation plan can be developed jointly with the Contracting Agency that addresses the consequences of the incompatibilities. This will enable adequate and appropriate risk provisioning to be made by all parties. Table 8.3, Table 8.4 and Table 8.5 give three examples of IRS entries which show how requirements can be handled. In these examples, the terms 'Aircraft Constraint' and Store Constraint' define the known capability (to the extent to which qualification evidence already exists) of the aircraft and the store.

Table 8.3   Example of a non-conflicting requirement

5.2.2.1 Altitude
Requirement Number: 0002
Aircraft Requirement: The aircraft shall have a store carriage altitude of 40 000 ft
Store Requirement: The weapon shall have a carriage altitude of 50 000 ft
Aircraft Constraint: Aircraft with store can achieve an altitude of 40 000 ft
Store Constraint: Store is qualified to 45 000 ft
Agreed IRS Requirement: The aircraft and the store in combination shall be compatible with carriage at 40 000 ft

Table 8.4   Example of a conflicting requirement resolved by relaxation

5.2.2.1 Altitude
Requirement Number: 0002
Aircraft Requirement: The aircraft shall have a store carriage altitude of 40 000 ft
Store Requirement: The weapon shall have a carriage altitude of 50 000 ft
Aircraft Constraint: Aircraft with weapon can only achieve an altitude of 35 000 ft
Store Constraint: Store is qualified to 45 000 ft
Agreed IRS Requirement: The aircraft and the store in combination shall be compatible with carriage at 35 000 ft

Table 8.5 Example of a conflicting requirement where additional work is required

5.2.2.1 Altitude
Requirement Number: 0002
Aircraft Requirement: The aircraft shall have a store carriage altitude of 40 000 ft
Store Requirement: The weapon shall have a carriage altitude of 50 000 ft
Aircraft Constraint: Aircraft with weapon can only achieve an altitude of 40 000 ft
Store Constraint: Store is qualified to 35 000 ft
Agreed IRS Requirement: The aircraft and the store in combination shall be compatible with carriage at 40 000 ft

Table 8.6 Example of a derived requirement

8.3.3.4 Digital Data Bus
Requirement Number: 0003
Aircraft Requirement: None
Store Requirement: None
Aircraft Constraint: MUX Bus voltage is below minimum specified by MIL-STD-1760E at the ASI
Store Constraint: Store provides a fully compliant MIL-STD-1760E interface
Agreed Integrated Product Requirement: The integrated product shall operate without changes to either the store or the aircraft

Table 8.3 identifies that although the store does not meet its specification, the performance of the aircraft and the store combination provides an acceptable level of performance. Therefore, no additional work is required to provide the capability.

Table 8.4 identifies that although the desired capability is constrained by the performance of the aircraft, following a review of the Concept of Operation, a relaxation of the requirement can be accepted by the Contracting Agency. This pragmatic approach means that no additional work is required to provide the capability, thereby reducing integration risk and therefore controlling the cost of integration.

Table 8.5 identifies that although the desired capability is constrained by the performance of the weapon, a review of the Concept of Operation means that although a relaxation of the requirement can be accepted by the Contracting Agency, a relaxation to the level with which the weapon can be compliant is unacceptable. The agreed requirement that the aircraft and the store in combination be compatible with carriage at 40 000 ft means that additional work is required by the WDO to provide the level of capability needed.

The nature of the additional work can therefore be agreed, and the relevant costs apportioned in the integration contract.

IRS requirements are derived from the needs of the Contracting Agency. However, it must be recognised that the Contracting Agency's requirements may not necessarily identify all the requirements needed for a successful integration programme. Therefore, there will be a need to derive additional requirements that do not flow from the Statement of Requirements. An example of such a requirement is shown in Table 8.6.

In this example, there will be a need to undertake representative system testing to prove that the aircraft constraint does not impact the performance of the store. This would be acknowledged as a programme risk and would need to be mitigated as early as possible during the integration contract. However, recognising a risk means that the cost of testing and the associated implications for the programme timescales can be addressed and the risk can be adequately managed from the outset.

By both DOs understanding the essential requirements, the aircraft and store designs can be progressed. The actual interface between the aircraft and the store will be progressively documented in the ICD with sufficient detail of both the aircraft and the store included to enable efficient integration to take place. However, for off-the-shelf stores, it is acknowledged that much of the ICD content may be defined at the start of the integration programme. The ICD will develop continuously throughout the integration programme and will only be finalised at the end of the integration programme. Throughout, it will be necessary to check the ICD for consistency with the IRS.

With regards to the qualification and certification activities, an IRS Validation Plan (IRS-VP) will be developed, and this should be directly traceable back to the requirements captured in the IRS. The IRS-VP will identify how the implementation of IRS requirements are proven (e.g. by test, analysis, etc.) and which party is responsible for the proving activity. The IRS-VP must be jointly agreed by the ADO and WDO and reviewed by the Contracting Agency but may not be a contractual document. The IRS-VP will enable the Contracting Agency to align the ADO and WDO integration activities with the wider Integrated Test, Evaluation and Acceptance (ITEA) planning with third parties (e.g. independent assessors). In order to bound the integration programme, the IRS-VP should be agreed by both the ADO and the WDO prior to accepting any development or integration contracts from the Contracting Agency.

The IRS Compliancy Matrix (IRS-CM) will be developed from the IRS-VP and will define the compliance/non-compliance against each IRS requirement. The IRS-CM will be jointly agreed by the ADO and the WDO and will contain evidence (i.e. references to reports where the detailed evidence is located) and, from this, will identify any limitations/agreed deviations or any IRS requirements that are deemed (and agreed) to be 'no longer required'. Where deviations are identified against the agreed IRS, these will be jointly reviewed by the ADO, WDO and the Contracting Agency so that timely action can be taken to understand the implications of the IRS deviation and determine the way forward. Responsibility for populating the IRS-CM will be programme dependent and will lie with whichever party is responsible for satisfying the proving requirement.

The integration risk management process described earlier will ensure integration risk is identified early, mitigated wherever possible before the integration contract is placed by the Contracting Agency and residual risks adequately managed. This focus will directly translate into efficient engineering activities with nugatory work being eliminated, leading to a successful integration programme.

Further Reading

Society of Automotive Engineers. (2004) Aircraft/Store Common Interface Control Document Format Standard. *Aerospace Standard AS5609*, Society of Automotive Engineers, Warrendale.

9

A Weapons Integration Scenario

9.1 Chapter Summary

Having covered the various aspects of aircraft systems integration of air-launched weapons, this chapter will draw together all the proceeding chapters and consider how they would contribute to a 'real' weapons integration programme (albeit, based on a hypothetical weapon). The chapter identifies example specifications and features for some aspects of the weapon and the aircraft which will be used to illustrate elements of the systems integration, by considering the activities on the left-hand side of the V-Diagram and the WBS. In order to aid the understanding of the scenario, a typical weapon loading (to the aircraft) to dispersion (safe separation from the aircraft) sequence will also be discussed.

9.2 Introduction

The previous chapters of this book have looked at a number of important elements for the integration of air-launched weapons. Whilst an understanding of the various aircraft and weapon subsystems is a prerequisite for the ability to integrate a new store, there is a far greater level of engineering capability required to link all the relevant factors together into an integrated aircraft and store product. This chapter and Chapter 10 will set the previous chapters into context by considering a weapons integration scenario. To assist in this, an imaginary weapon will be defined along with an aircraft system architecture which will have to control the new weapon. This chapter will explore the aspects of systems integration of an air-launched weapon by considering aspects of the left-hand side of the V-Diagram (the design activities) and the WBS defined in Chapter 2 whilst considering a weapon loading to dispersion sequence.

Aircraft Systems Integration of Air-Launched Weapons, First Edition. Keith A. Rigby.
© 2013 John Wiley & Sons, Ltd. Published 2013 by John Wiley & Sons, Ltd.

This will enable the interaction between the various aircraft subsystems to be explored and to discuss some aspects which will have implications for the interfaces between them.

By also considering the loading to dispersion sequence and the associated systems implications, this chapter will largely cover the left-hand side of the V-Diagram and some of the pervasive aspects such as safety. The right-hand side of the V-Diagram (the proving activities and the associated parts of the WBS) will be covered in Chapter 10.

9.3   The Weapons Integration Scenario

In this scenario, a new precision-guided weapon known as the XPGB is to be integrated with an aircraft. Table 9.1 defines the XPGB weapon capabilities (albeit, greatly simplified for the benefit of this example) covering the basic requirements identified in Chapter 1 and other important details which are relevant to its integration with the aircraft.

In this scenario, the XPGB will be integrated with an aircraft which has already been designed for the integration of smart weapons. The aircraft is a single-seat, aerodynamically unstable light combat aircraft with a complex Flight Control System. The pilot controls all weapon selections including target selection. The aircraft has five weapon stations: two on each wing and one on the fuselage centreline. The centreline pylon is dedicated to the carriage of a fuel tank to extend the radius of action of the aircraft. Each wing pylon has a store carriage mass capability of 750 kg. By employing a multi-weapon carrier with an empty mass of 80 kg capable of the carriage and release of two smart weapons, the total weapon load-out will consist of eight XPGB weapons.

The aircraft system architecture is defined in Figure 9.1 (based on the architecture described in Chapter 4).

The system has an identifiable SMS as part of the wider Armament Control System. The SMS contains a mix of both safety-critical and mission-critical functionality, but there is also a separate Mission Computer which provides all functionality associated with weapon aiming and the display of data to the pilot. The SMS is connected to the Mission Computer via the MIL-STD-1553B Avionics Data Bus and the SMP is the BC for the MIL-STD-1760 Weapons Data Bus. The avionics system also includes a combined Laser Inertial Navigation System/GPS (LINS/GPS) unit which is connected to both the Avionics and Weapons data buses and provides a raw GPS RF feed to the SMP. The SMP also includes the relevant switching circuits which buffer and distribute the GPS RF signals to each weapon station (for clarity, the figure shows only a single weapon station). Finally, each store pylon has its own WSSU which provides the control and monitoring of the pylon S&RE and the switching/routing of the MIL-STD-1760 power supplies and discrete signals.

Loaded to each of the four wing-mounted store stations are smart multi-weapon carriers (MIL-STD-1760 Carriage Stores) which contain a single MIL-STD-1553B RT connection with the MIL-STD-1760 ASI and provide two CSSI for weapon control. Each smart carrier contains its own electronics unit which acts as an RT on the aircraft Weapons Data Bus and provides a BC function to control the two XPGBs loaded on the carrier. It should be noted that if multi-weapon carriers have to be integrated as a part of the programme to integrate the XPGB, then this is in reality two integration programmes which are undertaken simultaneously. However, for this example, considerations for the control of multi-weapon carriers are covered separately (see Section 9.8.13).

Table 9.1 XPGB requirements

Requirement	XPGB capability
Lethality	250 kg HE Penetrator warhead with a programmable fuze which can be set via the weapon interface with the aircraft
Precision	3 m CEP achieved through a combination of GPS-assisted inertial navigation plus an IIR seeker for terminal guidance
Stand-off range	30 nautical miles when launched from 20 000 ft and deploying a set of range extendable surfaces (wings and stabilisers)
Store mass	The weapon consists of the warhead, a tail kit including the guidance electronics, a nose-mounted IIR sensor and a mid-body, deployable range extension (wing) kit. Total store mass is 310 kg
Mechanical attachments	Bail Lugs compliant with STANAG 3726
Electrical interface	MIL-STD-1760 Class 2 interface
Power supply for logic functions	28VDC1 only
High bandwidth signal requirements	HB1 Type B for receiving GPS RF signals from the aircraft
Data interface	MIL-STD-1760 multiplex data bus complying with MIL-STD-1553B
Maximum 'power-on' time	120 s at maximum environmental temperature (note that real weapons are unlikely to have such a constraint. This limit has been included in this example to illustrate a number of points later)
Minimum release height above terrain	300 m (to enable fuze arming to take place)
Control surface deployment	Spring-loaded surfaces that are locked in a stowed position until a physical safety pin is removed during the launch sequence
Terminal seeker	IIR seeker located in the nose of the weapon
Launch mode	Downward ejected (maximum ejection rate of 10 m/s)
Arming strategy	Smart fuze which detects two environmental changes; disconnection from the aircraft (via monitoring disconnection of the MIL-STD-1760 interface Address lines) and weapon confirmed as being under control of its auto-pilot within a timed window 5 s after disconnection from the aircraft (the timer is implemented in simple, and therefore, easily analysable electronic hardware)
Required data parameters from launch aircraft	• Navigation data: aircraft position (latitude and longitude) and altitude above mean sea level; aircraft velocity in three directions (north, east and down); GPS almanac data, UTC time (with a resolution of 64 µs); aircraft roll, pitch and true heading • Target position data: GPS co-ordinates (latitude and longitude) and target height above mean sea level • Lever arm data (X, Y and Z axes) and static offset in pitch, roll and yaw • Target confirmation data: image for scene matching and relevant meta-data • BIT initialisation data • Weapon selection data: Transfer Align select Note: aircraft data parameters for latitude, longitude, true heading, velocities (north, east and down), altitude, pitch, roll and yaw must be time-tagged with the time at which data was sensed by the aircraft system
Data parameters provided to the launch aircraft	• BIT status • Transfer Alignment status • Read back of data received from the aircraft (e.g. fuze settings)

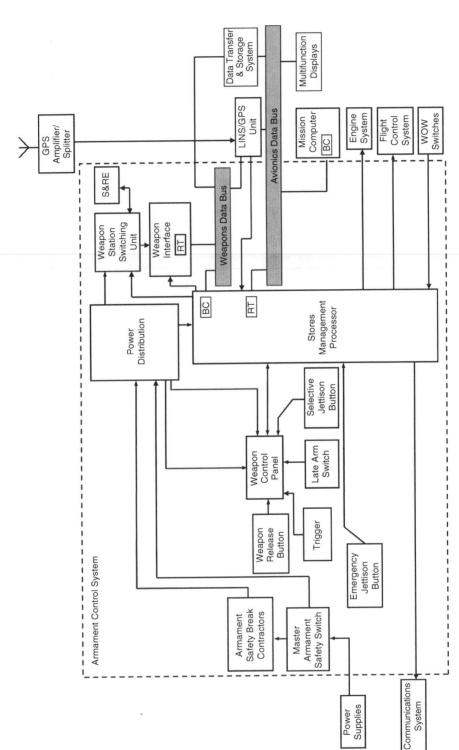

Figure 9.1 Aircraft system architecture

9.4   The V-Diagram Revisited

Figure 9.2 shows the final version of the systems engineering V-Diagram discussed in Chapter 3.

Although Table 9.1 defined some basic requirements for the XPGB weapon, these do not directly relate to either the Statement of Requirements or the content of the IRS as required for the highest levels of the systems process. The Contracting Agency will define the operational needs (derived from an Operational Analysis of how the integrated aircraft and weapons are to be used). These could be pitched in terms of the target set to be engaged (including the required fuzing mode selections to be programmed via the interface with the aircraft); the stores carriage configurations required with particular regard to the definition of mixed configurations of the XPGB with other stores (e.g. fuel tanks, Electronic Countermeasure pod, self-defence missiles); the aircraft operating envelope to satisfy operational needs when carrying the XPGB; the weapon release envelope (in terms of maximum and minimum release altitudes and speeds); target engagement/weapon aiming modes (such as pre-planned, target of opportunity, etc.).

In analysing the Contracting Agency's Statement of Requirements, the ADO will also engage with the WDO in order to gain an understanding of the intended operation of

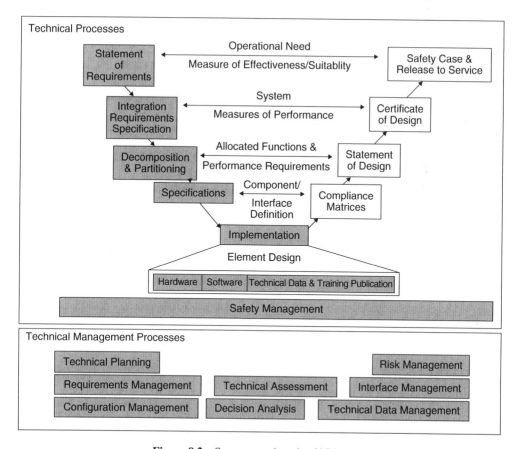

Figure 9.2   Systems engineering V-Diagram

the weapon and its various modes (e.g. terminal guidance needs). In effect, working with the Contracting Agency and the WDO, the ADO will need to determine the required System Design Requirements and the Performance Specification for the integrated product.

If the weapon has already been developed, then the master (ideal) weapon ICD will also need to be obtained from the WDO and analysed. Armed with the knowledge of the design of the weapon and the Contracting Agency's Statement of Requirements, the ADO and WDO can jointly develop the IRS and the associated verification methods (Qualification Plan). In compiling the IRS, the ADO and WDO will need to develop a set of integration requirements which satisfy the Statement of Requirements, taking into account any limitations which may need to be agreed due to any incompatibilities identified during the analysis of the aircraft and weapon designs (as discussed in Chapter 8). Whilst the Statement of Requirements defines the capability which the Contracting Agency requires industry to deliver, there is an intellectual activity required to translate these needs into a consistent set of weapon system integration requirements which form the basis of the integration programme. Time spent in undertaking this activity will have a significant benefit in identifying risks. As noted in Chapter 8, if the IRS is agreed by the ADO, the WDO and the Contracting Agency, the number of unpleasant surprises during the integration programme will be reduced.

9.5   Systems Integration Activities

Having agreed the IRS for the integration of the XPGB, there are a host of other activities which together take the IRS requirements through systems implementation and certification to an in-service capability. From the systems integration viewpoint, there are many activities as identified in Figure 9.3 (the WBS repeated from Chapter 2, where the white boxes identify systems integration aspects) which must be undertaken to deliver the in-service capability. Some activities relate only to the systems elements of integration, but some activities which fall largely into the aeromechanical domain do have some relationship with the modification and re-certification of the aircraft subsystems.

During the remainder of this chapter and in Chapter 10, the systems integration-related elements of the WBS as they could relate to the integration of a weapon such as XPGB will be discussed. The WBS defines all the activities that are required to discharge a successful weapons integration programme. However, the activities will be undertaken in a structured way in line with the systems integration process defined by the V-Diagram. In order to structure the considerations of the WBS activities, it is useful to understand to which parts of the integration life cycle they relate. The WBS can therefore be mapped to the V-Diagram as shown in Table 9.2. It should be noted that the Technical Management Processes in the V-Diagram do not have a mapping to the WBS. This is because the management processes are needed to underpin the integration programme and do not, in themselves, provide a direct contribution to the systems and aeromechanical integration activities. The exception to this is the Interface Management Process.

The remainder of this chapter will consider the WBS activities which relate to the design activities whilst Chapter 10 will discuss the proving activities. However, one set of activities which cover the whole life cycle and hence stretch across both sides of the V-Diagram is safety. As it is essential that the safety activities are commenced very early in the integration

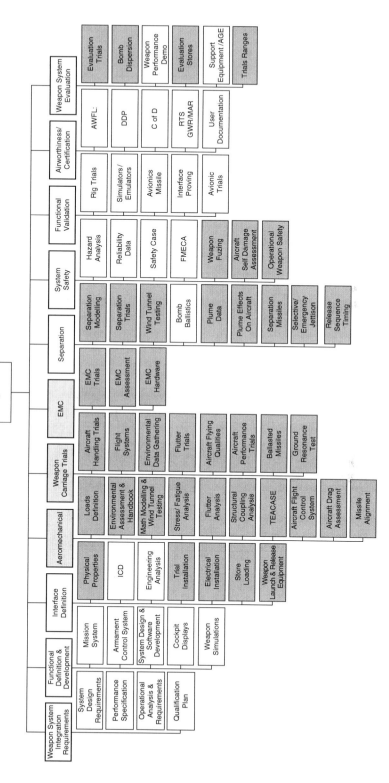

Figure 9.3 Weapons integration WBS

Table 9.2   WBS to V-Diagram mapping

Systems engineering activity phase	WBS elements
Statement of Requirements	Weapon System Integration Requirements
Integration Requirements Specification	Weapon System Integration Requirements
Decomposition and Partitioning	Functional Definition and Development
Specifications	Functional Definition and Development
	Interface Definition
Implementation	Functional Definition and Development
	Interface Definition
Compliance Matrices	Functional Validation
Statement of Design	Airworthiness/Certification
Certificate of Design	Airworthiness/Certification
Safety Case and Release to Service	System Safety
	Airworthiness/Certification
Safety Management	System Safety

programme, Section 9.6 will discuss some of the safety consideration that relate to the aircraft subsystem design activities.

9.6   Safety

For a properly designed aircraft system, safety would be an intrinsic property that would have been considered from the outset of the programme. In considering system safety, it is necessary to understand the hazards that need to be protected against. Table 9.3 defines a subset of typical hazards that are relevant when considering the integration of a new weapon and considers some of the aircraft subsystem design issues. Whilst this subset of hazards would have been considered and adequately mitigated during the initial design of the aircraft system, whenever a new weapon is to be integrated, then it is good practice to revisit these hazards in order to understand if there are any implications for the integration of the new weapon.

The overall Safety Programme Plan will define exactly which safety activities will be undertaken as a part of the integration programme. This will include the need to fully analyse each hazard to the point where it is adequately mitigated. This will generally be achieved when the HRI (see Table 2.3) has been reduced to an acceptable level. Whilst some hazards can be mitigated quickly, especially if they have already been addressed either in the design of the aircraft system or during the integration of a previous store, others may drive more detailed analyses which lead to system changes. From Table 9.3, the hazards that would require a greater level of analysis for the XPGB integration are 'extended application of power to weapon', 'incorrect transmission of store type or store presence to Flight Control System', 'crew release weapon below a safe height', 'weapons released at less than safe minimum interval' and 'armed stores present on aircraft when not required'.

The aircraft system considerations for these hazards when integrating the XPGB weapon could be as follows.

Table 9.3 Typical hazards and considerations during the integration programme

Hazard	Considerations when integrating a new weapon
Extended application of power to weapon	Due to issues with heat dissipation, the weapon has a maximum time which it can be powered. If powered for a longer period, damage to the weapon may occur. Whilst this is a hazard that relates to the weapon, there are also potentially serious considerations when the weapon is loaded on the aircraft. Over-heating of power supply components could lead to hot surfaces or even burning, which, depending on the installation on the aircraft, could have varying consequences. This hazard should be reviewed in line with the Bow-Tie model depicted in Figure 2.2 and the risk determined. This will enable a suitable method of mitigation to be considered, which could, for example, include changes to the aircraft power control system to disconnect power to the weapon station after a pre-determined maximum period. However, MIL-STD-1760 states that the application of 28VDC1 shall not degrade the safety of the weapon. Clearly, if exceeding the maximum power-on time of the weapon could cause safety issues, XPGB would therefore be non-compliant with the standard.
Incorrect transmission of store type or store presence to Flight Control System	For an aircraft which is aerodynamically unstable, the actual stores configuration loaded at any instant may be critical to continued safe flight. If this is the case, then whenever a new weapon is to be integrated, it will be necessary to understand how its mass, centre of gravity, roll inertias and so on could have an impact. Whilst the aircraft's Flight Control System would already have been designed to take account of this important feature of flight dynamics, any impacts caused by the new store should be considered.
Crew release weapon below a safe height	A weapon may, post release from the aircraft, have to deploy control surfaces or ignite a rocket motor. Depending on the aerodynamic stability of the weapon during release, deployment of the control surfaces or motor firing may need to be delayed to enable a level of aerodynamic stability to be achieved. Clearly, under the influence of gravity, the weapon will begin to fall, which means that there will be a minimum release height. Should the weapon be released below this minimum height above the ground, then it could impact terrain and break up causing debris. In the worst case, a weapon with a rocket motor may be damaged by the initial impact but could still fire its motor leading to an unpredictable trajectory, which, in turn, could lead to a high-energy impact with the launch aircraft. Whilst this is a hazard which could be directly caused by the action of the crew (i.e. releasing the weapon at an unsafe height above ground), it may be decided that this is insufficient mitigation (crew action would be the fourth priority mitigation action as defined in Table 2.4), and as such, a different method of mitigation chosen (e.g. provide a visible warning to the crew such that they are made aware that releasing the weapon when the aircraft is below the safe height is hazardous)

(continued)

Table 9.3   (*continued*)

Hazard	Considerations when integrating a new weapon
Weapons released at less than safe minimum interval	When undertaking multiple releases of a weapon, there will be a need for the ADO to understand the aerodynamic performance of the weapon during release to ensure that weapons released from adjacent stations do not collide. This is a particular problem for un-guided weapons where accurate impact spacing on the ground could be important. For guided weapons, release intervals are not usually critical and so it is usually a design decision for the ADO to determine the minimum safe interval from aerodynamic modelling. Minimum safe intervals are a property of the combination of the aircraft and the weapon and, as such, would not normally be defined in the weapon's specification
Weapon control surfaces deployed early such that surfaces strike aircraft or other stores	If a weapon has deployable surfaces such as wings or stabilisers which are moved from a stowed position to the free-flight position post release, then there is a hazard that should these be deployed too early in the aircraft/store separation cycle, they could impact the aircraft or other stores and cause serious damage. Although the weapon will have design features which should guard against the hazard becoming a consequential risk, the mechanism for doing this needs to be reviewed by the ADO. Should operation be initiated by the removal of a safety pin that is attached to a lanyard which is attached to the aircraft, such that it is not removed until the weapon is at a specific distance below the aircraft (the distance being set by the length of the lanyard), then the risk should be relatively easy to analyse. For example, the cause could be either the pin suffers fatigue and breaks, the lanyard has been made too short and so on. For each, a probability of occurrence could be calculated and suitable mitigation planned. If however, the deployment of surfaces is mechanised by the weapon post release, for example, by being electrically actuated after a separation delay, there could be several potential failure modes which need to be analysed. Areas for consideration would include the method in which separation is detected by the weapon, the method in which the post-separation delay is measured and the method of switching power to the electrical actuators. If software plays any part in the deployment of weapon control surfaces, then the ADO should be seeking to understand the integrity vested by the WDO in the software. Whilst from a pure weapon viewpoint, this may not be deemed to be a significant problem, and therefore software of an inadequate integrity level has been employed, when the weapon is integrated with an aircraft this could become a significant safety issue that will need additional analysis by the WDO. Whilst such analysis may not be performed until after the integration programme has been contracted, it is issues such as these which drive the need for safety to be carefully considered at the IRS stage
Weapon/debris strikes aircraft	Prior to release, some weapons may need to be prepared by ejecting turbo jet intake or terminal sensor covers. Whilst it will

(*continued*)

Table 9.3 (*continued*)

Hazard	Considerations when integrating a new weapon
	be the aircraft which should be responsible for initiating the removal of such covers (this should be clearly defined in the ICD as a part of the release timelines, etc.), the debris could be hazardous to the launch aircraft (or even the over-flown population). Whilst not necessarily a systems integration issue, there will be the need for aerodynamic modelling to be performed to understand the trajectory/spread of debris and an assessment made on the implications for the aircraft. For example, does the debris impact any sensitive areas such as external fuel tanks loaded on other stations on the aircraft? Such hazards can be very difficult to mitigate and so needs to be considered at the IRS stage (in the worst case, this could be a fundamental issue for the integration programme and may mean that the weapon can only be carried on certain stations, thereby reducing the weapon load-out and therefore the military capability of a single aircraft
Stores fail to release	Whilst the aircraft system should have been designed with a strategy for handling stores which fail to release (e.g. remove the weapon from the stores inventory but keep it available for jettison), any new weapon that is integrated needs to be assessed on its own state should it fail to separate from the aircraft when intended. A weapon with a turbo jet may need to have its engine started prior to release so that its internal power is available immediately on release. If however, the weapon contains a rocket motor which has failed to fire, then the reason could be due to a mis-fire. In this case, there could be a delayed ignition of the motor propellants leading to the weapon 'firing' whilst it has been re-locked onto the weapon station. Even for an ejected weapon, there could be a thermal battery that has been initiated prior to the intended launch. Whilst it is expected that the weapon will be designed to cope with a failure to separate from the aircraft, there may be aircraft system consideration for maintaining aircraft safety. This could, for example, include removing the weapon from both the release and jettison inventories. Also the handling of the aircraft when it lands will need to be considered. In such a case, it would be normal for the aircraft to be moved to a safe area whilst, for example, the weapon's thermal batteries discharge and cool down before special unloading procedures are implemented
Armed stores present on aircraft when not required	The store design should be such that it cannot arm itself until a time after it has physically separated from the aircraft. However, it will be necessary for the ADO to understand the arming strategy for the weapon such that it can be confirmed that it cannot be armed in the vicinity of the aircraft. If it is found that the weapon is able to be armed in the vicinity of the aircraft, then a hazard analysis of the weapon will be required (probably undertaken by the WDO at the request of the ADO) to understand the risk. Understanding the risk will enable mitigation to be defined

9.6.1   Aircraft/System Hazards

9.6.1.1   Extended Application of Power to Weapon

From Table 9.1, the XPGB has a maximum time which power can be applied of 120 s. This is at the maximum ambient environmental temperature, so it is likely that at a lower temperature, this time may be increased. However, the pilot may not know the exact ambient temperature which the weapon is experiencing, so from a weapons integration viewpoint, this time would be taken as the effective limit.

The XPGB will require power to be applied to enable BIT to be performed, to enable target data loading and to align its navigation system prior to launch. The weapon will have been designed such that all these activities can be accomplished within 120 s. Also, if an attack is aborted, then there will be a need to return the weapon to a safe power-off state. This cycling of power could be left to the pilot, but during times of heavy workload, a level of automation will be required such that power is applied when needed and removed before the 120 s limit is reached (it should be noted that a weapon design with a maximum power-on time would probably not be operationally usable and, therefore, would be unlikely to be designed with such a limit. However, this has been introduced in this example to highlight how deviations from the ideal can have a significant impact on the activities required to be undertaken during the integration programme).

Power may be switched to the weapon whenever the XPGB is entered into a valid weapons package by making the appropriate selections on the WCP (see Figure 9.1). The WCP is connected to the SMP which in this system architecture is the main processor in the Armament Control System. It is the SMP which will initiate the switching of power to each of the weapons selected in the weapons package, and therefore, functionality will be required within the SMP to ensure that power is only ever applied to the XPGB for a maximum of 120 s.

The effects of applying power for greater than the 120 s limit would need to be investigated. If overheating within the weapon could occur which could lead to a catastrophic failure (the WDO will have undertaken the relevant analysis during the development of the weapon to understand the effects of the time limit being exceeded), then there will be a need for the ADO to ensure high-integrity control of power to the weapon. Whilst the SMP may well be implemented in high-integrity hardware and software (to satisfy the 'no single failure criteria'), there may also be a need to ensure that power switching is also implemented in a high-integrity fashion such that there is sufficient confidence in the ability of the aircraft to remove power after a maximum of 120 s (as noted in Table 9.3, this contravenes the requirements of MIL-STD-1760).

If, however, exceeding the maximum power-on time only results in a weapon failure that is deemed to have a severity of 'marginal' or 'negligible' (see Table 2.1), then the system implementation of power switching could be simplified. Clearly, a simplified power switching architecture would be preferred to reduce the cost of system design. However, this can only be considered as a solution if the analysis indicates that this does not lead to an unacceptable safety issue for the aircraft and weapon combination.

9.6.1.2   Incorrect Transmission of Store Type or Store Presence to Flight Control System

As the aircraft in this example is deemed to be aerodynamically unstable, the Flight Control System may have to account for the effects of total store mass, effects on the overall aircraft's centre of gravity and any inertial effects created by stores carriage. Should the Flight Control

System have incorrect stores inventory data, this could lead to under- or over-corrections to the flight dynamics which could lead to loss of the aircraft. The aircraft systems will have been designed such that stores inventory data can be passed from the SMP to the Flight Control System (see Figure 9.1). However, the introduction of the XPGB into the inventory will require an assessment to be made of the implications for Flight Control System operation. Depending on the flexibility of the Flight Control System to accept varying store masses across different stations, there may not be any significant impact. However, to confirm this, aerodynamic modelling will be required covering all potential configurations which include the XPGB (including, e.g. when there is only one weapon fitted to a multi-weapon carrier). With regards to the aircraft systems, the SMP will need to be able to pass the relevant store parameter data to the Flight Control System and the Flight Control System may have to be modified if the XPGB stores configurations take the flight dynamics beyond the limits of the control laws already designed into the system.

9.6.1.3   Crew Release Weapon below a Safe Height

In this example, the XPGB has a minimum release height above the terrain of 300 m. From the weapon's viewpoint, this is the minimum height that the weapon will fall under the influence of gravity before fuze arming is achieved. From the aircraft's viewpoint, the greatest threat could be damage caused by high-velocity blast debris impacting the aircraft (due to the weapon arming just before it impacts the ground following a release below the safe height above terrain) thereby causing a consequence of catastrophic severity. A weapon impacting the terrain before it has armed may only be deemed to be mission-critical, unless the weapon disintegrates and debris rebounds to impact the aircraft.

In order to overcome this, the aircraft system could be designed such that whenever the aircraft is below a safe weapon release height, the release of an XPGB is inhibited. However, there are many system implications for this mechanisation. Whilst it may be the safety-critical SMS which inhibits the release, it will have to use data which is generated in potentially mission-critical sub-systems (e.g. in the LINS/GPS unit in Figure 9.1). Such data will have its own integrity level which will be dictated by both the available sensors, other hardware in the unit and the software controlling the unit/generating the aircraft altitude data. As such, it is unlikely (and probably undesirable from a system cost view) that the aircraft will have a high-integrity LINS/GPS unit.

An alternative mechanisation could be for the aircraft to employ a radar altimeter to accurately measure height above terrain. Although this too is unlikely to be a safety-critical piece of equipment, if used as a second input to the SMS (possibly via the Mission Computer), a strategy of comparing inputs from dissimilar sources could be used to determine the actual height above terrain. Ideally, a third independent input would be available so that a '2 out of 3' voting scheme could be used to discern if one of the sources was in error. Significant analysis of the failure rates of the different inputs would therefore be required to determine if the desired integrity can be achieved.

Assuming there are sufficient independent sources of height above terrain and that the combined integrity of the sources is deemed acceptable, then the SMS could be used to inhibit XPGB release below a minimum safe height. This could, however, also lead to a very complex and therefore costly system which would be exacerbated should a new height source also need to be integrated as a part of the XPGB integration programme. It is for this reason that a number of system trades will be required to investigate alternative solutions. With reference to

Table 2.4, designing for minimum risk (the ideal option) would be likely to drive the need for a modification to the XPGB weapon. If XPGB has completed its development, then implementing a potentially expensive mitigating design change to the weapon is unlikely to be palatable to the Contracting Agency. The next priority option (again from Table 2.4) of incorporating safety devices is in effect the aircraft system solution outlined earlier. For this integration example, the most cost-effective method of mitigating the hazard could be for the height to be displayed to the pilot with a warning symbol (possibly flashing display symbology on the HUD) when the minimum safe release height is transgressed. However, even such a warning display may not be adequate due to the integrity of the software and hardware driving the HUD. There could therefore be a need for the pilot to cross-monitor the HUD with other sources of data (e.g. a direct display of height data from a Radar Altimeter or a visual check of the outside terrain). The pilot would therefore need to be trained on the action to take when the warning is displayed. With reference to Table 2.4, this solution would be a combination of providing a warning device and implementing adequate training of personnel (the pilot). Before this course of action could be taken, the agreement of the Contracting Agency would be sought. If this is deemed by the Contracting Agency to be unacceptable, then it may be possible to explore changes to the desired in-service use of the XPGB to avoid low-level releases altogether. Clearly, if considered soon enough in the integration programme, it could potentially be traded when developing the IRS.

9.6.1.4   Weapons Released at Less Than Safe Minimum Interval

When considering the integration of a new weapon, there will be a need to ensure that the interval between consecutive weapon releases is considered. Ideally, an aircraft would not have any constraint placed on it, but aerodynamic realities can lead to stores colliding if they are released at too short an interval. Ballistic bombs particularly are prone to post release collisions in the vicinity of the aircraft due to the desire to have minimum impact spacing around the target resulting in ever shorter release intervals. For missiles which are forward fired by the ignition of a rocket motor, very short release intervals could be achieved if the time taken for the missile to clear the rail is short. However, lateral separation of missiles and post-launch manoeuvres will also have an influence on the achievable release interval. When determining the release interval, the implications of firing a missile that has a slow-burn motor followed by a missile with a fast-burn motor (where the second missile could feasibly catch the first missile in the vicinity of the aircraft) would need to be considered.

Store collisions could damage a weapon meaning that it does not leave the vicinity of the aircraft as expected or is unpredictable in its flight path. Either could cause the weapon to impact the aircraft with potentially catastrophic consequences.

To overcome this, aerodynamic modelling or wind tunnel testing will be required to determine the safe release interval. The theoretical minimum safe release interval will depend not only on the stores being released but also on what other stores are present on the aircraft which could affect the air-flows around the aircraft. In reality, a single minimum safe interval will be determined that is a compromise between an interval that is as small as possible and one which will always be safe, regardless of other stores present on adjacent stations.

As the minimum safe release interval is deemed to be safety-critical, the SMS is required to generate the store releases at the correct intervals. For ballistic bombs, where short intervals are desirable, several actual safe intervals could potentially be programmed into the SMP covering different store load-outs. However, this would significantly increase the proving activities on the right-hand side of the V-Diagram. For smart weapons which need to be launched in the LAR in order to achieve a successful target engagement, the practical release intervals can be increased to avoid problems with store collisions. Therefore, it is not uncommon for a smart weapon (such as the XPGB in this example) to be released at intervals of around 500 ms. For a free-fall ballistic bomb type, typical release intervals could be significantly less than 100 ms.

In a real system implementation, the release command could be generated by a Mission Computer (CCRP, CCIP, or an in-LAR flag). Generally, the Mission Computer software responsible for generating the release cues would not necessarily be of a high-integrity design. Therefore, in order to mitigate systematic failures which generate cues at an unsafe interval, the high-integrity SMS would act as a guard, ensuring that for a specific weapon type, even if cues are generated by the Mission Computer at unsafe intervals, weapons will not be released at less than the minimum safe interval.

9.6.1.5 Armed Stores Present on Aircraft When Not Required

It is assumed in this example that the aircraft SMS has been designed and certified to satisfy the 'no single failure' criteria defined in Chapter 4 such that the weapon cannot be armed before release nor prevent the weapon being made safe after having been selected for arming. From Table 9.1, the weapon employs a process to determine that it has separated from the aircraft, detecting two separate environments (disconnection from the aircraft and a timed manoeuvre under the control of the weapon's Flight Control System). During the design and proving of the XPGB, the WDO will have implemented a safety programme to ensure that all weapon hazards are adequately mitigated (albeit, not necessarily covering all factors caused by carriage and release from an aircraft). Additionally, during the integration programme, there will be a need for the ADO to examine the evidence to show that the weapon design is such that it cannot be armed until it has safely separated from the aircraft (in effect to understand the arming mechanism as defined in Table 9.1).

As the XPGB employs the MIL-STD-1760 interface, any pre-arming requirement would be controlled by employing the Store Control and Store Monitor protocols (as discussed in Chapter 6).

9.6.2   Weapon Hazards

The previous paragraphs have considered hazards from the aircraft viewpoint. But there is also a need for the ADO to understand all relevant weapon hazards and their mitigation. For example, if the weapon has a rocket motor, this could be hazardous to the aircraft if it ignited when not demanded. In such a case, the aircraft may not have deliberately powered the weapon, but due to some outside influence (e.g. electromagnetic interference), the rocket motor ignites. Whilst many of the obvious hazards will have been mitigated by the WDO, adopting similar methods as the ADO to mitigate aircraft hazards, there is a key difference between aircraft level hazards and weapon hazards. That is, a weapon is potentially in its least-safe state when it is undergoing a significant event such as a rocket motor being intentionally fired during the launch sequence.

Whilst a rocket motor will normally be relatively benign, at the point at which it is initiated, its propellant begins to rapidly burn, building up high-pressure gases (which are vented through the rocket motor nozzle producing thrust). Ignition creates a rapid rise in pressure within the rocket motor casing, which could rupture the casing at any weak spots causing the weapon to explode on the aircraft – clearly a failure of catastrophic severity.

Table 2.2 relates hazard probability of occurrence categories against the time of exposure to the risk. Whilst this gives a useful means of assessing the implications of risks of continuously operating systems (such as an SMS), it does not account for the risk attributed to a planned event such as firing of a rocket motor. Fortunately, the WDO will have assessed this risk and should be able to provide a justifiable probability of a rocket motor exploding on ignition. However, it will be the responsibility of the ADO to understand the associated additional risk to the aircraft and to ensure that adequate mitigation is in place which will have to be fully justified in the safety case for the aircraft/weapon integrated product.

Safety is the most important part of an integration programme and one of the most complex aspects, not least because the safety programmes of the ADO and the WDO need to be combined to ensure an overall acceptable level of safety is achieved. For this reason, it is recommended that the safety programme is initiated early in the programme so that the IRS can capture specific safety requirements which may then be agreed with the Contracting Agency to ensure an integration programme that is safe and has minimum technical risk.

9.7   Systems Requirements Decomposition, Design and Implementation

The following sub-sections give an overview of some relevant considerations when agreeing and decomposing requirements and the implementation considerations for the XPGB integration programme. Each subsection will consider the WBS activities from a functional implementation viewpoint.

9.7.1   Weapon System Integration Requirement

As defined in Chapter 2, the Statement of Requirements will be defined around the operational needs of the end user. The IRS needs to capture all relevant requirements which will be derived from the joint knowledge and understanding of the ADO, WDO and the Contracting Agency. This may include the output from an Operational Analysis of the intended use of the aircraft and weapon combination and requirements developed by the Contracting Agency in the form of System Design Requirements and Performance Specifications. For all requirements, there is the need to develop a Qualification Plan. The systems considerations for these parts of the WBS relating to the XPGB integration programme have largely been discussed in Section 9.4.

9.7.2   Functional Definition and Development/Interface Definition

Once the weapons system integration requirements have been defined and the IRS for the XPGB integration agreed, there will be the need to decompose and partition the requirements to the underlying system architecture. In order to do this, the requirements will need to be analysed taking due regard of the initial safety requirements and the contents of the XPGB ideal ICD and will need to consider the typical functions that have to be implemented/modified to integrate the XPGB.

9.7.3   Weapon Interfacing

The XPGB employs a MIL-STD-1760 Class 2 interface. As the aircraft has already provisioned an ASI at each of the store stations to which the XPGB is to be loaded, the wiring, switching and so on, will already be available and will therefore not normally require any changes for the integration of the weapon. With the MIL-STD-1760 infrastructure already implemented on the aircraft, only the logical functions and data transmitted over the data bus between the aircraft and the weapon will be affected.

The logical interface will be defined in the weapon ICD; for an existing weapon, this will be the ideal interface that is required for the weapon to achieve its specified performance. However, the ADO will usually negotiate the actual ICD that the aircraft and weapon will comply with. The intent will always be for the aircraft to satisfy the weapon ICD wherever practicable. However, in order to control the cost of the integration programme, there is also a need to minimise changes to the aircraft's functional architecture. The adoption of standard hardware and electrical interfaces obviously goes some way to achieving this, but this does not cover the logical data set required by the weapon. The logical part of the ICD therefore needs to compromise between meeting the idealised weapon ICD and minimising system changes to the aircraft, commensurate with the agreed (with the Contracting Agency as a part of the IRS) performance of the aircraft/weapon combination.

Whilst MIL-STD-1760 defines a number of standard messages, there will be a large number of data messages which are specific to the weapon. On the assumption that there are no data items which cannot already be sourced within the aircraft's subsystems, the message set can be agreed. However, where the negotiation of the ICD can be more difficult is when the resolution of specific parameters and/or parameter update rates needs to be agreed. Whilst the aircraft may be capable of transmitting data items at the rate required by the weapon, the ideal weapon performance is unlikely to be achieved unless the update rate (i.e. the rate at which the data parameters are produced) and accuracy of the data as generated by the aircraft's subsystems satisfy the demands of the weapon.

Agreeing the functional element of the ICD is likely to be a protracted process, as experience shows that even when the ICD is agreed, once the integrated system is tested, it is likely that errors and misunderstandings in the ICD definition will be discovered, requiring clarification of the ICD and, possibly, further system changes to be implemented. For this reason, it is essential that significant effort is expended early in the integration programme to ensure that the ICD is correct in order to alleviate the risk of expensive system/weapon changes later in the programme.

The functional element of the ICD defines the interface provided by the aircraft with the XPGB. In this example, this is the data bus transactions on the Weapons Data Bus. The aircraft system will also have a definition of the internal system interfaces, for example, the Avionics Data Bus ICD. The ADO will need to define all the changes to these internal ICDs, which in turn, help to define the changes that are required to individual subsystems. These subsystem changes can then be implemented as design changes in individual equipment and software items. With reference to Figure 9.1, the integration of XPGB is likely to drive changes to the Weapons Data Bus ICD, the Avionics Data Bus ICD, the SMP software, the Mission Computer software and possibly the Flight Control Computers (if the mass, centre of gravity and inertia

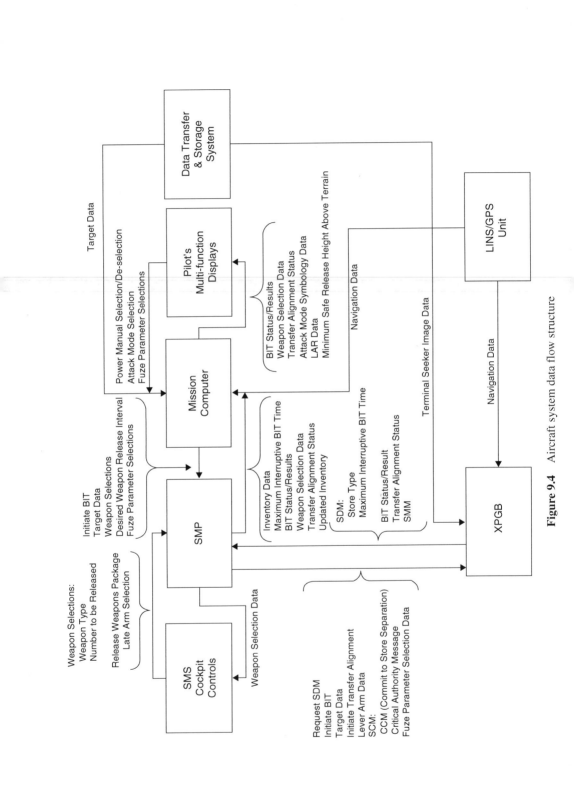

Figure 9.4 Aircraft system data flow structure

of XPGB are significantly different to existing integrated stores). It is assumed that there is no change to the logical sequence of events to prime and release the weapon (use of MASS, Late Arm, release button, etc.), although the sequence may need to also account for any weapon-specific interlocks in its priming sequence (e.g. a weapon that has not achieved an acceptable level of Transfer Alignment may be removed from the weapon package).

The final elements that need to be implemented are the timelines for XPGB priming and release. Whilst, in this example, the functionality will primarily reside within the SMP, there will also be consequential changes required to other subsystems such as the Mission Computer in order for the data which support the required state transitions of the timelines to be available to the various subsystems when required. This will include implementing functionality to measure the time the weapon has been powered and either to turn it off when the XPGB 120 s limit is reached, or to generate the appropriate display cues to the pilot. The change to the aircraft subsystems will also need to ensure that the instantaneous height above terrain is made available such that if the minimum XPGB release height (300 m) limit is breached, either release is inhibited or the pilot is notified (whichever implementation has been determined as a result of the safety analysis).

9.7.4 Data Flows between Aircraft Subsystems

Whenever a new store has to be integrated with an aircraft, its data interface requirements will have to be satisfied by the aircraft system. This is defined in the ideal ICD, which will have been derived from the design implementation needed to satisfy the customer's requirements/ concept of use for the store. The ICD defines all the required data items which need to be exchanged between the aircraft and the store, their accuracy and update rates and so on. The ADO will therefore need to determine how the data requirements will impact on the existing design of the aircraft system. However, it should be noted that the data provided at the aircraft/ store interface will be sourced/used by various aircraft subsystems, with each subsystem being responsible for parts of the store control functionality. The subsystems will need to communicate with each other so the aircraft system will have a number of internal interfaces that are used for passing data between its subsystems. These internal interfaces will be documented in their own ICDs which are controlled by the ADO (but may also include sub-agreements with the DOs for the various pieces of equipment/software).

The aircraft system will therefore have a defined distribution of software functions (to individual subsystems and equipment). In order to control the XPGB, various data will need to be passed between these software functions. The simplified data flow structure for XPGB integration is as shown in Figure 9.4; the need for and use of these data items will be discussed in the following sections. When partitioning new functionality to existing aircraft subsystems, it is important to understand the various data exchanges required to control the new weapon as this will drive changes to subsystem/software specifications.

9.8 Loading to Dispersion Sequence

In order to decompose the system requirement and to partition these to the system architecture, the ADO will need to understand the entire sequence of operation of the aircraft and weapon from loading the weapon to the aircraft through to safe separation of the weapon from

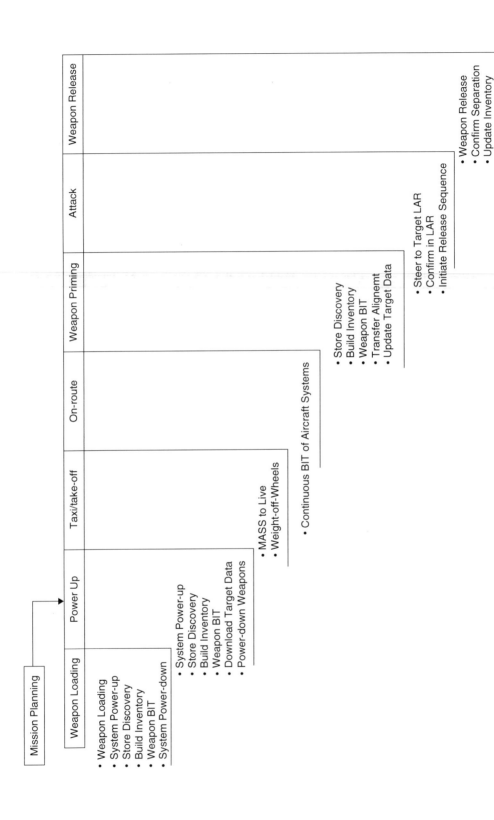

Figure 9.5 Loading to dispersion sequence

the aircraft. Mission planning will also have to be considered as this will be the source of the data required to identify to the aircraft systems where the targets are and provide target engagement data for the weapon. In the case of the XPGB, which uses images of the target to match with those detected by the IIR sensor, this will also include the relevant image files and any meta-data required by the weapon to correlate the planned image with the detected image.

A key part of analysing the loading to dispersion sequence is to gain an understanding of the integration requirements and to help the weapons integrator to develop the architectural changes required to the aircraft systems. This will entail determining how the functionality is to be partitioned across the aircraft subsystems and, from this, to understand the various data exchanges that will be required between those subsystems to enable an optimised control architecture for the weapon to be defined.

Figure 9.5 shows the various phases of the mission from weapon loading and mission planning, through pre-flight activities (power-up and taxi/take-off) and on-route activities, to the end game phases (weapon priming, attack and weapon release). A typical sequence of actions (system power-up, store discovery, build inventory, etc.) for each phase is also identified. Much of this information will be determined from the knowledge and experience gained from existing aircraft operations, an examination of the system requirements and the functionality of the weapon. This knowledge provides the basis on which to develop the decomposition of requirements and the partitioning of lower-level requirements to subsystems, equipment and software. The following sections will explore some of the systems integration considerations relating to the partitioning of the XPGB functionality to the aircraft system defined in Figure 9.1.

9.8.1 Weapon Loading

Prior to loading the weapon to the aircraft, there will be a need for the armourers to ensure that the aircraft is safe and that there is no power present at the electrical interfaces. Once this has been confirmed (usually with the aid of test equipment), the weapon will be moved into position under the S&RE and raised to a point where the store is close enough for the electrical interface connector to be mated with the MSI. Other connections such as electrical fuzing leads and the mechanical lanyards that will be used to deploy the XPGB's control surfaces post release will be rigged. When all connections to the weapon have been made, the weapon will be raised such that the S&RE hooks can engage with the weapon's bale lugs and the weapon can be locked in place. The S&RE sway braces will then be tightened to the appropriate torque setting.

9.8.2 System Power-Up/Store Discovery

Following loading, the relevant parts of the aircraft system will be powered up. From Figure 9.1, this would probably include the Armament Control System, the Mission Computer, and the Multi-function Displays. Figure 9.6 shows a timeline of the key activities.

The MASS will be set to the 'Standby' position thereby applying the logic supply to the SMP which will then initialise its hardware and software. As the XPGB has a MIL-STD-1760 interface, the SMP will request the WSSUs to interrogate the Interlock interfaces at each weapon station to determine if a MIL-STD-1760 store is loaded (note, in Figure 9.1, only a single station and its associated WSSU are shown). The status of each station's Interlock interface will

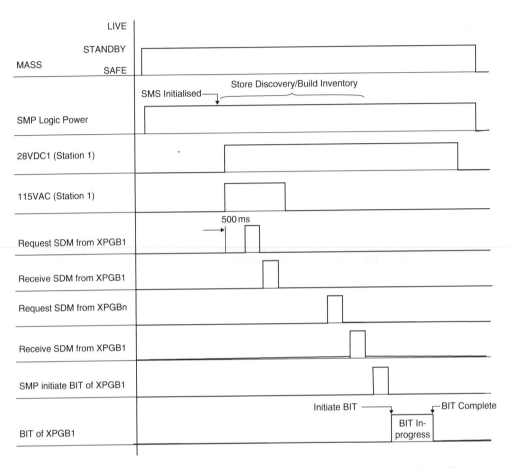

Figure 9.6   System power-up, store discovery, build inventory, and weapon BIT timeline

be reported to the SMP by the WSSUs, and the SMP will then command the application of 115VAC and 28VDC1 power supplies to the relevant stations. Both power supplies are applied initially until the aircraft has discovered exactly which store is loaded to the station. On the application of power, the stores loaded at these stations will begin their initialisation sequence.

9.8.3   Build Inventory

MIL-STD-1760 requires a Mission Store to be capable of responding to a valid request to transmit a SDM within 500 ms of power being applied. The SMP will request each of the stores (XPGB1 to XPGBn in Figure 9.6) in turn to transmit their SDMs which contain data identifying the store type and other data as defined by the standard which the aircraft may also use. As a part of the integration programme, the SMP will need to be modified to recognise the XPGB type and to introduce the relevant new control functionality for the weapon.

Once all stations have been interrogated, the SMP will build a picture of the stores loaded to the aircraft and to which stations they are loaded. Recognising the XPGBs, the SMP functionality will identify that the weapons only utilise the 28VDC1 power supply. The SMP will therefore command the WSSUs to remove 115VAC power from all stations where XPGBs are loaded.

The SMP will then distribute the inventory data to other subsystems as required. In the Weapon Loading Phase for this example, this will be to the Mission Computer only (as the Flight Control System has not been powered up).

9.8.4 Weapon BIT/System Power-Down

In this example (due to a previous design decision), the Mission Computer requires inventory information so that store BIT can be initiated automatically when stores are initialised. BIT status and results are displayed on the pilot's Multi-function Displays. For the XPGB, it should be noted that when the weapon is initially powered, it will undertake initialisation tests to ensure that it is safe, these tests being completed before the SDM is provided to the aircraft. Having received the inventory data from the SMP, the Mission Computer will request the SMP to schedule BIT for all the XPGBs. This request will require specific messages on the Avionics Data Bus (a BC to RT message transfer).

The XPGB defines, via the SDM (as provisioned by MIL-STD-1760), the maximum time it will take to perform BIT. As different stores will take different amounts of time to complete BIT, the SMP may pass these data to the Mission Computer so that it can configure its store BIT functions specifically for the XPGB. Should a weapon remain in BIT longer than this time, the SMP will determine that the weapon in unserviceable and will return this status to the Mission Computer.

The XPGB ICD will define the data transactions and timings required to initiate BIT, for BIT to run and then for the XPGB to report its status. As shown in Figure 9.6, this will be controlled by the SMP, scheduling the relevant transactions at the correct time. These data exchanges will require the system integrator to define any additional messages that need to be introduced into the Weapons Data Bus transaction tables and the action to be taken to an incorrect or no response from the XPGB.

Once the XPGB has completed its BIT, it will make the results available to the SMP which will upload the data and then (in this example) power down the stores by commanding the WSSUs to remove the 28VDC1 power from the relevant store stations. The SMP will then transfer the BIT information via the Avionics Data Bus to the Mission Computer, which will display the results on the Multi-function Displays. Should any of the weapons have returned a 'failed' BIT status, then these weapons would normally be removed from the available weapons inventory. However, depending on the nature of the failure, some level of weapon performance may still be available, albeit degraded. Whether the weapons are retained for employment or not will depend on the capabilities of the actual weapon and the rules of engagement for use of the weapon.

BIT should test as many of the weapon's subsystems as is feasible without degrading safety (e.g. it is not possible to test devices containing pyrotechnics for their function – the best that can be achieved is to test that initiator squibs are connected and have a resistance which is within specification). To fully test some weapon subsystems, for example, a GPS receiver or IMU, then there may be a need to route GPS RF signals to the receiver subsystem or to Transfer Align the IMU (Transfer Alignment is covered in Section 9.8.7).

During the Weapon Loading Phase, these actions are undertaken to ensure that all the stores that have been loaded by the armourers are serviceable and that the aircraft umbilical cable has been correctly mated with the MSI.

Once loading is complete, the operator will power down the aircraft system (for the SMP, this will be achieved by setting the MASS to the 'Safe' position) and ensure that all weapons are safe (safety pins fitted to S&RE, etc.). The aircraft may then be left for some time until the sortie is to be flown.

9.8.5 Download Target Data/Power-Down Weapons

In preparation for the sortie, the aircraft and weapons will be powered, the inventory built and BIT performed (as shown in Figure 9.6). The next significant action (notwithstanding that actions such as Transfer Alignment may also be performed during this phase but are not shown in order to simplify the example) is to select the weapons into the attack package and to download target data into each of the XPGBs.

In this example, the weapon selections are performed by the pilot making the selections (e.g. weapon type, number to be released, etc.) on the Weapon Control Panel. The SMP will transmit the relevant data available to the Mission Computer over the Avionics Data Bus, and in turn, the Mission Computer will pass this to the Multi-function Displays.

Target data will be developed in the Mission Planning System and will consist of co-ordinates defining the location of the target, specific attack parameters that the weapon may need and fuze settings data. For complex weapons that are able to manoeuvre, the data may also contain routing and waypoint information which the weapon will use to determine the flight path to the target to be followed post release. The data generated by the Mission Planning System will also include data which the aircraft systems will use to route the aircraft to the launch point and may include data defining the coefficients for the XPGB LAR, which will be used by the Mission Computer to configure its LAR algorithm. All the data will be compiled into a MiDEF file for transport to the aircraft.

For TOO, these data will be generated by the pilot using data generated by the aircraft sensors (which will have been used to geo-locate the target).

The aircraft's Data Transfer and Storage System (see Figure 9.1) is connected to both the Avionics Data Bus and will download the complete MiDEF file into the Mission Computer. The Mission Computer will interpret the file and distribute the data as required. Any data destined for the weapons will be transferred over the Avionics Data Bus to the SMP. The SMP will then schedule the data to each of the XPGBs in accordance with the ICD. Depending on the amount of data required by each weapon, the data could be downloaded as part of the normal cyclical Weapons Data Bus transactions. However, as the XPGB requires an image file of the target to aid its terminal guidance, this will also have to be downloaded. The total amount of data required by an individual XPGB is likely to be very large to the extent that downloading using the normal cyclical transaction on a MIL-STD-1553B data bus would be operationally unacceptable and would likely exceed the maximum power-on time for the XPGB. Therefore, in this example, the normal bus activity would be suspended and the MIL-STD-1760 MDT protocol invoked to download the target data including the image file. For this reason, the Data Transfer and Storage System is connected to the Weapons Data Bus so that the image file can be downloaded to the weapons without overloading the Avionics Data Bus.

From Table 9.1, we know that the weapon has a fuze which is programmable by data transmitted from the aircraft over the MIL-STD-1760 interface. Assuming that the fuze can exploit the functions defined in the Fuze Data words defined by MIL-STD-1760, it is likely that the actual fuze function parameters would be selected from the pilot's Multi-function Displays and Mission Computer. If the Mission Computer and the displays did not already contain the relevant fuze selection options, the Mission Computer software would need to be modified to introduce this new capability. New Avionics Data Bus transactions may also need to be scheduled to pass the fuze selections from the Mission Computer to the SMP so that these can be downloaded to the XPGB over the Weapons Data Bus.

Once the target and fuzing data has been downloaded into all of the XPGBs, the SMP will command the WSSUs to power down the weapons. For this example, where the XPGB has a maximum power-on time, the full sequence of power-up downloading of data, reading back the data and checking its integrity will need to be completed within the 120 s limit.

9.8.6   Taxi/Take-Off/On-Route Phase

The XPGB power-on limit means that during aircraft taxi, take-off and the on-route portion of the sortie, the weapons will be powered down. Once all ground crew actions have been completed, the pilot will set the MASS (which would only have been selected to the 'Standby' position during the power-up phase – see Chapter 4) to the 'Live' position ready for flight. When the aircraft becomes airborne, the WOWs will transition to the flight state. Even with these two safety interlocks overridden, the Armament Control System will still be safe as the conditions to effect a release (which would include having a valid weapons package selected and the Late Arm switch selected) would still be in a state which provides an adequate level of integrity.

During the On-route Phase, the various aircraft equipment will be performing continuous BIT to ensure their continued serviceability. The SMP, which is likely to be implemented in high-integrity hardware and software and satisfying the 'no single failure' criteria defined in Chapter 4, is likely to have implemented strategies for graceful degradation. That is, the system will be able to withstand a level of degraded performance due to failures and remain safe and available to effect a weapons release. Continuous BIT will be the mechanism for monitoring the health of the SMP and to ensure its integrity is not compromised.

As the aircraft approaches the target area, the sortie will enter the Weapon Priming Phase where the weapons will be powered up and tested, before Transfer Alignment is initiated and the weapon selections confirmed and so on.

9.8.7   Weapon Selection and Priming

In this phase, the pilot will select the weapons package on the Weapon Control Panel. The SMP will then signal to the Mission Computer that a valid weapons package has been selected and will instruct the WSSUs to power up the relevant XPGBs. The SMP will follow the store discovery process outlined in Section 9.8.2 (and Figure 9.6) and check that the inventory is still as it was on the ground (i.e. the SMP will check that all weapons have initialised correctly and are responding to transactions on the Weapons Data Bus). As the pilot will only want to release serviceable weapons, there will be a need to view the XPGB BIT results on the relevant Multi-function Display page.

On successful completion of BIT, the SMP and the Mission Computer will have a shared view of the health status of each of the weapons in the package. If a weapon has failed BIT, then the Mission Computer could request the SMP to alter the weapons package such that only serviceable weapons are included, or the SMP, being informed of a failure, could power up a replacement XPGB. However, this will take some time to perform, and as the XPGB has a maximum power-on time, the strategy implemented in this example is for the aircraft system to power up all XPGBs, and then for the SMP to choose which will be employed to satisfy the number required by the selected weapons package. Any weapons which have failed BIT will automatically be removed from the release inventory by the SMP. This decision and an indication of which weapons have been selected will be communicated to the Mission Computer for display to the pilot.

With the serviceability of the weapons determined, Transfer Alignment can commence. The XPGB has both a GPS receiver and an IMU. Both will need to be initialised and the IMU will also need to be aligned to the aircraft axis system using a standard Transfer Alignment protocol controlled by the SMP. The SMP will switch the GPS RF signals routed from the aircraft's LINS/GPS Unit (which is connected to both the Avionics and Weapons data buses) to each of the XPGBs so that the weapons' GPS receivers can start to acquire satellites. As the LINS/GPS Unit is connected to the Weapons Data Bus, the SMP will need to schedule RT to RT message transfers to all XPGBs in order to transfer the navigation and Transfer Alignment data identified in Table 9.1. So that each weapon can convert the aircraft's Transfer Alignment Data into its own navigation co-ordinates, the SMP will also need to transfer lever arm data specific to each weapon. Once Transfer Alignment has commenced, the SMP will continue to schedule the relevant transactions and poll the quality status of each weapon's navigation solution. Once a weapon reaches an adequate level of alignment with the aircraft axis system, then the SMP will be notified by the Mission Computer that the weapon is ready for deployment.

Although the SMP may have access to the alignment status of each weapon, the pilot will also need to be provided with this information. Rules of engagement may dictate that the pilot cannot release weapons unless they are fully serviceable and have an acceptable navigation solution. Therefore, alignment status of each weapon will need to be transmitted as an RT to BC data transfer from the SMP to the Mission Computer so that the information can be further relayed to the pilot's Multi-function Displays. Should a weapon not achieve an acceptable level of Transfer Alignment, then the pilot may determine (after a period of time that will be defined in the aircrew manual) that there is a problem with the weapon. The rules of engagement may dictate that such weapons are manually deselected from the inventory. In such a case, the pilot will need the ability to power down selected weapons which fail to align (in reality, the pilot may recycle the power to a specific weapon in an attempt for it to correctly Transfer Align).

9.8.8   Update Target Data

Assuming that the sortie has been prepared such that a number of different targets can be attacked using the XPGB load-out, each weapon may have already been programmed pre-flight with the relevant data. During the Weapon Priming Phase, should one or more of the weapons become unserviceable, then there may be a need to understand the relative priority of the various targets such that the most important are attacked with the available weapons. Therefore, serviceable weapons may have to be re-programmed with new target data. This may also be required should the planned target for some weapons have changed during the aircraft's sortie (e.g. if the target has been destroyed by another means).

Re-programming of the weapons could be undertaken by the pilot; however, this will increase workload in the cockpit at a critical phase in the sortie. Remembering that the XPGB has a power-on limit that must be adhered to, a greater level of automation may be required. Here, the system designer would have to decide how such functionality is to be partitioned between the Mission Computer and the SMP. This decision will have a bearing on the Avionics Data Bus transactions between these two equipment (i.e. it will have an impact on the Avionics Data Bus ICD).

With the weapons aligned and loaded with the correct target data, the sortie will transition to the Attack Phase.

9.8.9   Steer to Target LAR/Confirm in LAR

Even without a propulsion source, the XPGB has a large maximum stand-off range. Whilst this is intended to allow the aircraft to launch the weapon from a safe location away from the intended target, there is a need to ensure that it is only launched if it has enough kinetic energy to reach its target; that is, the aircraft is within the LAR (the SMP may be designed to inhibit release of the XPGB if the aircraft is not in the LAR). In this example, the XPGB does not calculate its own LAR so the aircraft will need to do this. This will require a LAR algorithm to be implemented in the Mission Computer software and displayed to the pilot.

When the Attack Phase is entered, the pilot will select a system mode which presents the relevant symbology on the Multi-function Displays. The symbology will provide steering cues to guide the pilot (or the aircraft's auto-pilot) towards the LAR associated with the target. The Mission Computer will commence the calculation of the LAR and provide the relevant data over the Avionics Data Bus to the Multi-function Display. As noted in Chapter 5, this could be a relatively simple display with markers identifying LAR entry and exit points and cross-track limits in relation to the aircraft's current position (see Figure 5.3), or it could be a more complex polygon. The actual LAR display implementation is likely to be dictated by the processing capabilities of the Mission Computer and the display update rates of the Multi-function Displays. Therefore, in order to understand the optimum implementation of the LAR, an analysis will need to be undertaken to determine how much spare processing capacity there is within the Mission Computer, which will also need to include an analysis of the optimum update rate to the LAR display.

As the aircraft nears entering the LAR, the pilot will commit to the release of the weapons, with the aircraft system actually initiating the release once inside the LAR.

9.8.10   Initiate Release Sequence

Prior to the XPGB being released from the aircraft, there will be a sequence of operations required to prepare the weapon. The sequence of operations to prepare and release

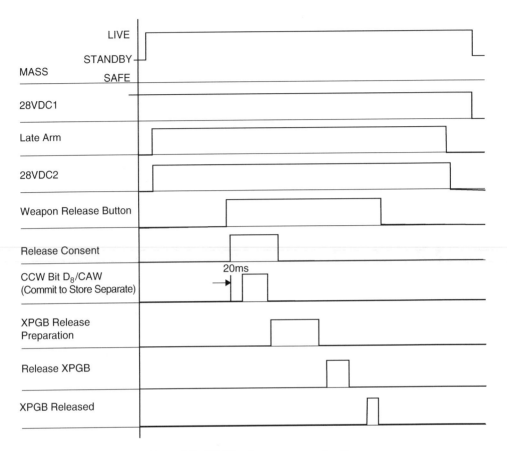

Figure 9.7   XPGB release sequence timeline

the XPGB is likely to be very much automated within the aircraft subsystems as timing tends to be critical. Figure 9.7 shows the weapon release sequence timeline for the XPGB.

Prior to releasing the weapons package, the pilot will select the MASS to the 'Live' position. Figure 9.7 shows that 28VDC1 is on as the weapon is already powered. As the aircraft nears entering the LAR, the pilot will select the Late Arm switch in the cockpit to on. In this example, this is the final interlock before the release actions for the weapon are performed. The status of the Late Arm switch (which interfaces with the WCP (see Figure 9.1)) is reported to the SMP. With all the interlocks removed, that is, the WOWs are signifying a flight state, a valid weapon package is selected, MASS is 'Live' (enabling Fire Supplies) and the Late Arm switch is on, the SMP will command the relevant WSSUs to turn on the MIL-STD-1760 28VDC2 power supply and the relevant S&RE fuzing supplies. The weapon will still be in a safe state at this point, but as the aircraft enters the LAR, the sortie transitions into the Weapons Release Phase.

9.8.11   Weapon Release Phase

9.8.11.1   Weapon Release Sequencing

When the weapon aiming solution indicated by the LAR is acceptable, the pilot will operate the Weapon Release Button which interfaces with the WCP. The WCP will report this action to the SMP which will initiate the sequence to release the XPGBs entered in the weapons package.

The SMP will command the WSSU to energise the MIL-STD-1760 Release Consent discrete signal for the first weapon in the release sequence. The standard defines a stabilisation time of 20 ms for Release Consent so that the weapon does not respond to transient signals. Once this stabilisation time has expired, the SMP will start to schedule the required transactions on the Weapons Data Bus. For this XPGB example, this is to schedule a Store Control Message (SCM) with the 'Commit to Store Separation' bit (bit D_8) set in Critical Control Word 1. The weapon, now in a phase where it is preparing to be released, will perform all the required protocol checks and then acknowledge receipt of a correctly formatted message using the Critical Monitor Word 1 (with the Committed to Store Separation bit set). The Critical Control Word 1 indicates to the store that it is about to be released and that it should perform the actions required so that the store will be self-sufficient post release. This could include the initiation of a thermal battery which would provide power to the weapon's electronics post release. The SMP will continue to poll the status of the weapon whilst it is performing its pre-release functions. Once the XPGB has performed its release preparation functions, it will signal this to the SMP using the Critical Monitor Word 1 in the SMM. The store is now ready to be released.

If at any time during this sequence of operations the weapon fails to respond as required (e.g. the demanded state changes from the weapon are not actioned within the allotted time), then the aircraft will instigate a failure management strategy. Generally, this will need to ensure that the weapon remains safe at all times but will also result in the weapon being deemed to have suffered a failure. Release of the weapon will therefore be terminated, power will be removed and the weapons release inventory updated accordingly (the failed weapon would be retained in the stores inventory so that the Flight Control System remains aware of its presence and the weapon remains available for jettison if required).

The XPGB is an ejected weapon, and this will be commanded by the SMP instructing the relevant WSSU to operate the S&RE. The WSSU will perform this action and then monitor the status of the S&RE for a 'store gone' indication. Once this has been detected, the SMP will be informed and the weapons inventory updated accordingly with the Mission Computer and the Flight Control System being notified of the new inventory.

Once the first XPGB has been released, the SMP will initiate an identical sequence to release the next weapon and then repeat the sequence until an attempt to release all weapons in the package has been completed. Should weapons fail to release (a hang-up has occurred), the inventory will still be updated, but the data provided to the Mission Computer and the Flight Control System will now differ. The Flight Control System will be given data identifying the presence of physical stores on the aircraft. The Mission Computer will be given data which indicates which weapons are still available for release (weapons that have hung or have failed BIT are normally only available for jettison).

Aerodynamic modelling will be undertaken to define the optimum release sequence and the minimum safe release intervals for stores configurations including XPGB. Again, the SMP will control the XPGB release sequence and timings and ensure that individual weapons are released when it is safe to do so. With a guided weapon such as XPGB, ballistic weapon aiming is not required (to deliver multiple weapons to an aim point becomes a function of the weapon rather than the accuracy of weapon aiming algorithms).

9.8.11.2 Weapon Separation

Safely separating the weapon from the aircraft either as an intended release or by jettison is largely the responsibility of the aerodynamics discipline. Safe separation of the XPGB is achieved when the weapon is ejected with sufficient downforce to overcome aerodynamic forces and at a time that separates it spatially from any other stores which are being released. The XPGB has the added complication that post release it will deploy its control surfaces including wings. Therefore, in analysing the release trajectory of the XPGB, it will be necessary to understand the influence of the aerodynamic conditions around the weapon at release and to identify any unwanted or erratic behaviour (e.g. weapon pitching up after release such that it could impact the airframe).

9.8.11.3 Bomb Ballistics

The XPGB is a guided weapon, although its initial trajectory before its flying surfaces are deployed will largely follow a ballistic path. Following deployment of the wings, lift will be generated meaning that the bomb can glide to its intended target.

For ballistic bombs, the trajectory will be very important as this will ultimately dictate where the weapon impacts the ground. As discussed in Chapter 5, there will be a need to understand the ballistic performance of the bomb so that the weapon aiming calculations can be optimised. A great deal of modelling will be required to understand the bomb trajectory for different environmental conditions such as temperature and air density. The results of the ballistic modelling will have to be coded into the Weapon Aiming Computer (the Mission Computer in Figure 9.1) either as complex equations or as look-up tables used in generating the release solution.

9.8.12   Selective/Emergency Jettison

The XPGB wing kit is deployed on release from the aircraft by removing a safety pin which is retained by a lanyard locked into an electrically operated fuzing unit in the aircraft pylon. When the XPGB is jettisoned, the fuzing unit will not be energised (weapons are not usually set to arm when they are jettisoned) and so the wing kit will not deploy. The weapon will therefore follow a pure ballistic trajectory.

If a single weapon station is selected for jettison, then the XPGB will separate safely with no risk of colliding with other stores. However, if multiple stations are selected for jettison or a 'clear-all stations' emergency jettison is initiated by the pilot, then there could be a number of stores being released in a very short time period. Indeed, in an emergency, there will be a need to clear stores from the aircraft safely in the shortest possible time.

Aerodynamic modelling will define the minimum safe intervals and sequence for jettison, and this could vary depending on exactly which stores are loaded on the aircraft. These sequences and step interval times will be encoded into the SMP which will manage the jettison of all stores.

9.8.13   Carriage Store Control

The maximum XPGB load-out is increased by the use of multi-weapon carriers. In this example, the aircraft has not previously employed these so the XPGB integration programme will also need to include the integration of the Carriage Stores.

For the purposes of this example, it is assumed that the multi-weapon carriers are MIL-STD-1760 Carriage Stores. The Carriage Stores provide a compliant CSI with the aircraft and contains electronics that are able to translate aircraft control signals and data for the control of two compliant CSSIs. This simplifies the integration of multiple weapons, but does place some demands on the aircraft subsystems, primarily in the SMP.

The introduction of the Carriage Store could mean that there is a separate DO responsible for the carrier. This means that there are likely to be at least two ICDs required: one for the aircraft/Carriage Store and one for the Carriage Store/XPGB. Whilst this does add organisational complexity, the interface management principles described in Chapter 8 still hold true. However, in this case, the relationships between the ADO, WDO and Contracting Agency now have to account for the addition of the Carriage Store DO (and may need to be considered when compiling and agreeing the IRS). The Carriage Store DO could be contracted as a separate entity within the programme or as a subcontractor to either the ADO or the WDO. However, consideration should be given to the through-life management of the aircraft where having the Carriage Store DO contracted by the ADO means that any future smart weapons integration programmes do not need to involve the XPGB WDO (as they would need to if the WDO had subcontracted the Carriage Store DO). Such an arrangement between the ADO and the Carriage Store DO would mean that the ADO has greater freedom of action for future integration programmes.

Once the contracting arrangements have been defined, the engineering activities can commence. When considering the aircraft system functionality, a number of issues are important as follows.

During the store discovery phase when the SMP is initialising the aircraft's stores inventory, it will detect that a Carriage Store is present. The aircraft will then need to request the Carriage Store to interrogate its own stores load-out and report what is loaded. The Carriage Store will employ the standard MIL-STD-1760 store discovery protocol (requesting the SDM from each XPGB) and report the relevant data to the aircraft. Once the SMP has compiled the data, the other aircraft subsystems (notably the Flight Control System and the Mission Computer) will be informed of the inventory. It should be noted that the aircraft will also initiate a BIT of the Carriage Store to ensure its serviceability.

The aircraft will still be able to communicate with individual XPGB weapons, but this will require the packaging of data destined for a particular weapon into a message transmitted to the Carriage Store. MIL-STD-1760 makes provision for this with a dedicated subaddress (07) for a 'Peeling' protocol (a protocol where the first word provides routing information for use by the Carriage Store. Software processing within the Carriage Store strips the routing information away (peels the word) and re-transmits the remaining data words to the intended Mission Store).

When operating weapons on a Carriage Store, there are several differences from when they are connected directly to the aircraft. The weapon lever arm data will be different as the Mission Stores are no longer located at the aircraft S&RE, but will have an additional displacement due to the Carriage Store. There will be additional delays in data transmission and receipt by the Mission Stores as the Carriage Store will have to process and re-transmit data. This will increase the staleness of the data. To overcome this, the time-tagging protocol defined by MIL-STD-1760 will need to be employed such that the XPGB can minimise the effects introduced by the Carriage Store. These additional issues are not insurmountable, but they do need to be considered at the outset of the programme.

9.8.14   Training Capability

Although the previous description has covered the integration of the physical XPGB weapon, there will also be a need to provide a training capability for the pilot when the real weapon is not fitted to the aircraft. In peace-time, the majority of flights will be with the aircraft system configured in a training mode. When the XPGB is integrated with the aircraft, the training modes will need to be reviewed and modified accordingly in order to provision a high-fidelity simulation. To achieve this, the SMP and Mission Computer software will be modified such that all the XPGB control functionality is simulated (including display formats and selection/switching sequences) but no safety-related discrete signals are generated. Expected weapon responses (such as indications that Transfer Alignment is complete) will have to be mechanised as a part of the XPGB simulation.

As the Mission Computer software in this example implements the LAR functionality, this can also be employed as a part of the training facility. If however, the weapon had been designed to calculate its own LAR and to provide this to the aircraft, it would mean that for the training mode, the aircraft would also have to implement a training LAR solution. This could be of a lower fidelity than the real weapon LAR (in order to reduce processing over-heads), although this could lead to slightly different cues to the pilot depending on whether a live or a simulated weapon is being operated.

By placing a large portion of the XPGB simulation in the SMP, the internal system inter-faces can also be simulated such that the remainder of the aircraft subsystems (e.g. Mission Computer) may only require relatively minor changes, employing much the same functional-ity as the live weapon. In doing so, the additional subsystem changes required to introduce an XPGB training facility are minimised.

Whilst the aircraft system may include a level of weapon simulation, there may also be a need for training weapons to be employed. These will be supplied by the WDO and should mimic the operational weapon completely. However, if these weapons are not intended to be released from the aircraft, the aircraft system design may need to account for this. One way of doing this for a weapon complying with the MIL-STD-1760 interface is to interrogate the SDM. If the weapon describes itself as a training variant, then the SMP would inhibit the release of the weapon. However, jettison may still be required in case of in-flight emergencies.

9.8.15   Implications of Aeromechanical Aspects – Weapon
Physical Alignment

There are areas of the aeromechanical aspects of a weapons integration programme which drive considerations for systems integration. One area that can have implications is the physi-cal position of the weapon. A weapon's navigation system will be aligned to the aircraft's navigation system and will therefore expect to experience identical perturbations to those transmitted by the aircraft (albeit displaced from the aircraft's datum and received a finite time after the perturbation was sensed by the aircraft).

Figure 9.8 ASRAAM missile launch from an aircraft wing station. Reproduced by permission of BAE Systems

Lever arm data is defined under static conditions. Under dynamic conditions such as during flight, aerodynamic forces will act to deflect the weapon from its static position, exploiting any elasticity in the aircraft S&RE. For store stations mounted on the aircraft fuselage, the degree of movement is likely to be small. However, for wing-mounted stations, there can be a great deal of movement caused by wing flexing.

The flexing of aircraft wings can be very complex consisting of a ripple along the wing and a leading edge to trailing edge twisting motion. The harmonic period of these deflections will differ, and this may lead to a very complex displacement of the weapon station in relation to the navigation datum of the aircraft.

For a short-range Air-to-Air missile that needs to be locked on to its target before launch, wing flexing can have a significant influence on the ability of the aircraft system to provide seeker pointing commands and for the missile seeker to respond. In the worst case, for an aircraft that is violently manoeuvring in order to get into a firing position, it could be almost impossible for the missile to lock on to its target. In order to understand if wing flexing could be a problem for missile aiming, a significant amount of aerodynamic modelling will be required along with performing wind tunnel testing. The flexing could be characterised and equations developed which provide an approximation of the actual wing flexing for various input conditions (Mach number, angle of attack, etc.). These equations would need to be implemented into the weapon aiming software for the missile so that some of the problems caused by wing flexing can be alleviated.

Figure 9.8 shows a typical weapon launch of a short-range Air-to-Air missile from the wing station on a Tornado aircraft (note also the camera pod mounted under the aircraft fuselage which is used to record the missile launch).

Further Reading

United States Department of Defense. (1978) Digital Time Division Command/Response Multiplex Data Bus. *Military Standard 1553B*, US Department of Defense, Philadelphia.

United States Department of Defense. (2004) Aircraft/Store Electrical Interconnection System. *Military Handbook 1760A*, US Department of Defense, Philadelphia.

United States Department of Defense. (2007) Aircraft/Store Electrical Interconnection System. *Military Standard 1760E*, US Department of Defense, Philadelphia.

United States Department of Defense. (2007) Mission Data Exchange Format, *Military Standard 3014*, US Department of Defense, Philadelphia.

10

A Weapons Integration Scenario: System Proving and Certification

10.1 Chapter Summary

Having used the weapon loading to dispersion sequence in Chapter 9 to discuss the systems integration activities in the left-hand side of the V-Diagram, this chapter will continue to draw together the proceeding chapters and expand on other areas of the integration programme, in particular, those activities relating to the right-hand side of the V-Diagram (system proving and certification).

10.2 Introduction

Once the system has been designed and implemented in the various aircraft subsystems, there is a need to progressively prove that requirements at each level on the left-hand side of the V-Diagram have been embodied correctly. In order to do this, there will be a need to gather evidence which can be used to certify the aircraft/weapon combination. In undertaking the system validation, verification and certification activities, various test assets will be employed such as weapon simulators and avionic weapons. The aim is to prove that the aircraft and the weapon have correctly implemented the ICD before undertaking activities which will test the functionality which is to be delivered to the end user. At each stage, evidence will be gathered to support the safety case and certification activities.

The following sections give an overview of relevant considerations during the system validation, verification and certification activities for the XPGB integration programme.

Aircraft Systems Integration of Air-Launched Weapons, First Edition. Keith A. Rigby.
© 2013 John Wiley & Sons, Ltd. Published 2013 by John Wiley & Sons, Ltd.

10.3   Simulators and Emulators

In the early stages of the integration programme, the ADO may have developed a very rudimentary simulation of the interface with the weapon based on an early version of the ICD. This could consist of simple switched discrete signals, indicator lights and a Data Bus Analyser programmed to respond to the transactions on the MIL-STD-1553 Weapons Data Bus. Whilst this level of simulation may be adequate in the early stages of SIL testing, when the aircraft system has been fully integrated, a higher fidelity simulation will be required in order to fully exercise the aircraft systems and to generate evidence which can be used to support the safety case and the eventual certification of the aircraft/weapon combination.

The higher fidelity simulation will be provided by the WDO. If the weapon is itself being developed, then a fully capable simulator may not be available. In such a case, the WDO will develop an emulation (known as an emulator) of the real weapon interface as an early interface proving tool ahead of a full-function simulator being available. The simulator will provide a fully representative, full-function simulation that provides the necessary flexibility to test the integrated system across the full range of the ICD. For example, the simulator may be programmable such that response times to data bus commands may be varied to explore how the aircraft and weapon can cope with varying levels of system performance.

10.4   Avionic Weapons

Understanding that the end product of the integration programme is the relevant clearance documentation and safety case for the aircraft/weapon combination, the data required to provide the necessary level of evidence can only be achieved by using fully representative system components for both the aircraft and the store. The use of a weapon simulator may not be deemed to be fully representative, and therefore, the ideal situation is to use real weapons in the final stages of SIL testing.

Whilst this is the ideal, health and safety regulations mean that weapons containing explosives cannot be used. Therefore, the WDO will build a special version of the operational weapon which has all the pyrotechnic devices removed. In their place, instrumentation may be fitted in order to be able to remotely monitor the internal states and data. This version of weapon primarily exists to support the integration testing in the SIL.

Avionics weapons may also be constructed to have their physical attributes such as mass and centre of gravity identical to the operational weapons and for them to be qualified for flight. This enables the avionic weapons to be fitted to the aircraft to support flight trails.

10.5   Interface Proving

The combination of emulators/simulators and avionic weapons are used to fully exercise the aircraft/store interface in the SIL. As noted earlier, all aspects of the interface need to be tested within the bounds of the ICD. However, the ICD is compiled to cover the interface between a single store and the aircraft. SIL testing will also need to cover the interface functionality when multiple stores are loaded to the aircraft. Clearly, for power supplies and discrete interface signals, this should not be a significant issue. However, the MIL-STD-1553 data bus requires special attention.

MIL-STD-1553 is based on a balanced transmission line (of 75 Ω impedance). The standard was originally designed for avionic systems where each data bus stub can always be terminated with the correct impedance (either an equipment is fitted, whereby the RT provides a load impedance on the bus, or an open circuit stub is terminated with a resistor of the correct value). The data bus system was not designed to have RTs which are disconnected from their stub during system operation. Of course, when the RT is located within a weapon, this is exactly what happens when the weapon is released.

SIL testing must therefore consider the effects on data bus operation when the stores configuration changes. Testing will need to be performed with various stations loaded with representative data bus impedances. With the aim of collecting evidence to support certification, there is a need to be able to load the data bus with multiple avionics weapons.

In this XPGB integration example, the aircraft is capable of operating up to eight weapons loaded on multi-weapon carriers. Practical realities of space available within the SIL may mean that it is not physically possible to have all weapons and carriers present. Therefore, in the XPGB case, where the weapon electronics are located within a tail kit, it would be a more practical solution just to use real tail kits rather than the full weapon (for fidelity of XPGB testing, employing a full weapon does not provide any benefits over using just a tail kit).

The SIL testing will generate a large amount of test data that will be used to confirm that all the aircraft system components are operating as designed and comply with the ICD. The WDO will need to provide documentation which identifies any deviations from the ICD and to assess the effect on weapon performance. To support this goal, a compliance matrix will be compiled against the ICD. Where there is a multi-weapon carrier involved in the integration (as there is for the XPGB programme), there will be multiple matrices.

10.6 Rig Trials

The changes required to individual equipment and software implemented at the bottom of the V-Diagram will be undertaken by several organisations such as the ADO, the SMP supplier organisation and possibly specialist military software houses. Each organisation will be responsible for ensuring that its system component meets its own specification as flowed down from the ADO before delivering it to the SIL. The SIL is the key facility for bringing together the various equipment and software and integrating them into the system.

The process of integration is a progressive process starting with confirming that the delivered component actually provides the required functionality. Initially, each component will be exercised on a rig which generates all required inputs and responds to all the required outputs using simulation. As each system component is proven, the simulation of its interfaces will be replaced by other real system components which themselves have undergone this first stage of testing on the rig. In an ideal world, testing would proceed as described earlier, and the systems development process would have assured that all system components work as required. In reality, at each stage problems will be identified and be investigated, the root cause identified and the problems with the system components rectified. This will require the systems engineering process to be iterated through all the affected levels.

As the overall system is progressively integrated, there will be a point, particularly relating to the weapon, where a higher fidelity simulation is required.

10.7   Avionic Trials

When there is sufficient confidence in the integrated functionality as tested in the SIL, testing will move to the aircraft. Whilst SIL testing can generate large amounts of certification data, flight testing is important in identifying emergent system properties when the integrated aircraft and store are operated in a representative environment by aircrew. Whilst flight testing is also required to determine the environment to which the store is exposed during flight, to demonstrate safe separation of the weapon from the aircraft during a release or jettison, and so on, testing of the aircraft system implementation in flight can identify the most complex of problems which are not apparent during SIL testing. Flight testing will also allow the pilot to explore the efficiency of operation of the weapon when there are other tasks to be performed relating to flying the aircraft, responding to defensive aids system alerts and so on.

Avionics weapons will be used extensively during flight trials, but due to their expense, they will not normally be released from the aircraft. The WDO may also have incorporated a telemetry system into the weapon in place of the warhead so that data in flight can be collected to support the certification of the weapon on the aircraft (from the viewpoint of the WDO).

10.8   Electromagnetic Compatibility

An aircraft has many sources of electromagnetic radiation, from generators of high electric field strengths (such as radios, radar, etc.) and magnetic fields (such as power cables), to lower power emissions caused by interference exported from signal cables and so on. During the design of the aircraft, great care is taken to control electromagnetic interference. Individual equipment will be specified to avoid the generation of unwanted levels of interference and also to be resistant to interference generated elsewhere. However, when considering the broad frequency spectrum and high levels of interference to which aircraft systems have to be immune, it is not unusual for individual equipment specifications to define challenging requirements. It is therefore common for equipment to be non-compliant against the full specification with a number of deviations being identified during the equipment's qualification programme.

In the worst case, equipment that deviates from its specification could malfunction when subjected to external interference or could cause other equipment to malfunction. For most equipment, the upset could be temporary and completely recoverable, albeit its function may be temporarily lost until the source of interference is removed. For a mission-critical piece of equipment, this could be an annoyance but, ultimately, may be accepted by the Contracting Agency. If the malfunction occurs frequently or causes the equipment function to be corrupted such that it needs to be restarted, then this could be unacceptable. However, the equipment supplier will undertake extensive testing to ensure the equipment is immune to interference at defined levels. The ADO will also undertake whole aircraft testing to determine that the complex electromagnetic environment which is generated when all system components are operating together does not impact safety or cause unwanted degradation in the operational capability.

Aircraft equipment which is deemed to contain safety-critical functionality (such as the Flight Control System or SMS) would normally be specified to be resistant to higher levels of interference to ensure a lower risk of malfunction. For the SMS, such a malfunction could result in the uncommanded release of stores. As this is a hazard that would be identified during the system design, the effects of electromagnetic interference will be one of the potential causes that would need to be considered and mitigated by the design solution.

Just as a great deal of effort goes into minimising the effects of electromagnetic interference on aircraft systems, the WDO will also ensure that the weapon can operate safely when exposed to its electromagnetic environment. Whilst during the testing phase of the weapon development programme the design can be proven to remain safe when exposed to a specified environment, this cannot account for the complex, multi-frequency fields generated when it is loaded to an aircraft; the electromagnetic environment generated by a specific aircraft type being unique. The compounding of the electromagnetic environment can be significantly more severe than the environment experienced during individual design proving tests. When a new store is integrated with an aircraft, there is the need to undertake a whole system test to ensure that the environment generated by the aircraft/store combination does not unduly affect the operation of either the aircraft or the store. As a worst case, the environment could cause a catastrophic initiation of pyrotechnic devices such as the rocket motor or warhead.

However, finding problems during on-aircraft testing could mean that changes are required to the aircraft, the weapon or both, in order to ensure that a safe combination can be fielded. This will drive cost and introduce programme delays.

Clearly, electromagnetic compatibility between the aircraft and weapon needs to be assessed early in the integration programme. This will be undertaken as an engineering assessment of the known environment generated by the aircraft and the levels to which the weapon has been previously tested. In doing so, the ADO will also need to assess if any interference generated by the weapon could interfere with aircraft systems.

Electromagnetic environmental modelling may also be employed in order to assess, in some detail, the predicted field strengths across the frequency spectrum. Any potential incompatibilities can then be investigated further and mitigation action taken. This could drive design changes to the aircraft or the weapon (although if a weapon is an existing inventory weapon, the Contracting Agency may not wish to consider modifying the entire stockpile of weapons if an aircraft palliative can be found). For urgent operational requirements, the Contracting Agency may accept the risk of a catastrophic weapon malfunction in order to field the capability to support on-going military operations. In such instances, aircraft palliatives could include restricting or disabling aircraft equipment which cause the unwanted interference with the weapon.

For a more detailed overview on the electromagnetic integration of aircraft systems, see the chapter on this subject in the Encyclopaedia of Aerospace Engineering by MacDiarmid (reference 10.1002/9780470686652.eae477). Also the requirement for the control of electromagnetic interference characteristics, as defined by MIL-STD-461, is another useful reference source.

10.9 Airworthiness and Certification

Once all testing (both within the SIL and on the aircraft in flight) has been completed and all changes required to correct problems have been embodied and tested, the aircraft/store combination will need to be formally cleared for entry into service with the end user. The documentation required to ultimately provide the Release to Service to the end user is progressively constructed, starting at the individual component level, through subsystem level, to the total aircraft system level. The following sections will outline some of the considerations for activities on the right-hand side of the V-Diagram for the XPGB integration programme.

10.10   Declaration of Design and Performance/Statement of Design

In integrating the XPGB with the aircraft, a number of system components will have been modified. Wherever system components are provided by subcontractors to the ADO, there will be a need for the individual organisation to provide their formal statement of compliance with the specification placed on them. The subcontractor will be required to make a statement of compliance against every requirement in the specification (sometimes referred to as a Declaration of Design and Performance (DDP)). Any deviations from the specification will also be detailed by the subcontractor in the DDP.

For the XPGB integration example, the SMP would be modified by the relevant subcontractor organisation. The DDP would be submitted showing that all specification requirements had been correctly implemented. However, if we assume that the original SMP had been developed with a deviation against the RT impedance such that instead of providing the 75 Ω equivalent impedance, the impedance is actually 65 Ω. This impedance mismatch will cause signal reflections on the data bus which will introduce additional noise and distortion onto the electrical signals being transmitted on the bus. In a worst-case scenario, this could actually lead to increased transmission errors, requiring multiple re-tries or, when the effects of un-terminated data bus stubs is taken into account (after weapons have been released), a complete failure of the data bus to communicate with the remaining weapons.

For the purposes of this example, it has been assumed that during the original SMP programme, this deviation would have been accepted by the ADO (as it was deemed not to impact the baseline system performance). For the XPGB integration, which it is assumed was achieved by a software modification, the deviation is carried forward unchanged.

During SIL testing, the MIL-STD-1553 data bus testing would have investigated the implications for the operation of the XPGB in the various stores configuration which the Contracting Agency has specified as being required in service. On-aircraft EMC testing would also have provided valuable evidence on the added effects of additional noise on the data bus. For the sake of this example, it is assumed that all this testing has been successful and no problems have been identified with the operation of XPGB on the aircraft.

The results of all the evidence relating to the Armament Control subsystem implementation will be captured in a SoD for the subsystem. This will include the SMP data bus impedance deviation and the rationale (i.e. the analysis of all the relevant test results) for accepting the deviation at the subsystem level.

All subsystems will have their own SoDs. Each SoD will make formal statements on the completeness of the subsystem implementation along with any areas where the implementation is non-compliant with the subsystem requirements. Such areas of non-compliance should be such that they only have minor effects on subsystem operation, otherwise some form of corrective action should have been taken to rectify the problem. An analysis of the non-compliant functionality will identify deviations or limitations which need to be recorded in the SoD.

10.11   Certificate of Design

The C of D draws all the individual SoDs together to take an aircraft-level view of the completeness of integration. At this stage of the airworthiness process, there will be a need to understand the effects on the end product of the interaction of all deviations. For example, the

SMP deviation may not cause the XPGB any problems, but the LINS/GPS Unit is also connected to the Weapons Data Bus. As this unit could be covered by a SoD from say, the Navigation subsystem, the C of D will need to assess if this has been adequately tested from an overall system viewpoint and to confirm that the Armament Control subsystem deviation does not have wider implications within the avionics system.

10.12   Safety Case

As discussed previously, the ADO and the WDO together are responsible for delivering a safe aircraft/weapon capability to the end user. The Safety Case is essential to achieving this. The Safety Case can take many forms, but the key attribute is that it documents how all hazards have been mitigated and captures the evidence generated from the design process, testing and analysis. The Safety Case will be developed with a level of independence from the design team and complements the C of D. Together, all the evidence is brought together into the Airworthiness Flight Limitations (AWFL) document.

10.13   Airworthiness Flight Limitations

The AWFL document draws together all data from the Safety Case, the C of D, flight testing, EMC testing, WDO documentation (for the XPGB) and so on. It defines from the ADO's viewpoint the safe limits for operation of the aircraft with all integrated stores including the XPGB. The AWFL is usually developed independently of the ADO design teams and the WDO such that a dispassionate view can be taken of any limitations identified during the integration programme. It is the AWFL which is the ADO's end deliverable from the integration programme to the Contracting Agency.

10.14   Release to Service

The Contracting Agency holds the final responsibility for delivering the aircraft/store combination to the end user. The Contracting Agency will combine the AWFL from the ADO with equivalent data from the WDO (GWR), in order to construct the Military Aircraft Release (MAR) documentation. Whilst the AWFL will have been compiled by the ADO from its knowledge and experience and test flying by experienced test pilots, the MAR may be further constrained to account for the skills of a typical squadron pilot and will combine this with the existing clearances for the aircraft. Once complete, it is the MAR which provides the formal RTS of the new combination of aircraft and store.

10.15   User Documentation

Although the MAR sets out the limit of operation of the aircraft, the end user will require a whole host of other documents (technical publications) which explain the details of aircraft operation. For example, there will need to be new publications identifying how to safely load the weapon to the aircraft, the aircrew manual will need to be updated to cover the operation of the new capability and there will need to be new documentation on how to handle and service the new weapon. Whilst both the ADO and WDO may provide many of these documents to the Contracting Agency, the end user may have their own documents which need to be updated with the relevant new information. Either the ADO or WDO could be contracted to do this or the end user may take on this task.

10.16   Weapon System Evaluation

Weapon System Evaluation (often called service evaluation) is normally led by the Contracting Agency, and the end user with support as required from the ADO and WDO. These activities are beyond the scope of the systems integration process so will not be covered in detail here.

Weapon System Evaluation is an activity that needs to be undertaken before the new capability is placed with the operational squadrons, and it is a means of not only checking that the delivered capability is as required, but is the first opportunity for the end user to exercise the capability in near-operational environments. The evaluation will explore and demonstrate the actual performance of the operational weapons in realistic scenarios operating against real targets (albeit, on a weapons test range – see Figure 10.1). The evaluation is also used to explore the use of test equipment and procedures required to support the operation of the weapon in service.

Figure 10.1   Missile launch against a real target. Reproduced by permission of BAE Systems

10.17 Conclusion

Chapters 9 and 10 have used a simplified and, in some instances, a contrived example of integrating the imaginary XPGB with an aircraft. However, in doing so, many of the considerations to which the weapons integrator will be exposed have been discussed. Whilst this gives a flavour of some of the potential pitfalls, a real integration programme will throw up many more issues almost on a daily basis, which will need to be carefully managed by the ADO, the WDO and the Contracting Agency.

Further Reading

MacDiarmid, I. (2010) Electromagnetic integration of aircraft systems, in *Encyclopaedia of Aerospace Engineering*. John Wiley & Sons, Ltd, Chichester.

United States Department of Defense. (2007) Aircraft/Store Electrical Interconnection System. *Military Standard 1760E*, US Department of Defense, Philadelphia.

United States Department of Defense. (1978) Military Standard 1553B: Digital Time Division Command/Response Multiplex Data Bus, US Department of Defense, Philadelphia.

United States Department of Defense. (1999) Military Standard 461: Requirements for the Control of Electromagnetic Interference Characteristics of Subsystems and Equipment, US Department of Defense, Philadelphia.

11

Introduction to 'Plug and Play' Weapons Integration

11.1 Chapter Summary

In an ideal world, it would be possible to bring together a weapon and aircraft that have no knowledge of each other. Both would employ standard mechanical attachments and a standard electrical interface. The aircraft/weapon combination would also have a standard method for both sides of the interface to discover each other and re-configure their operation such that the aircraft is able to exploit the capabilities of the weapon with no system or software changes. Re-configuration would be achieved by configuration data only. If this could be achieved, we would have what is referred to as 'Plug and Play' weapons integration. In 1998, NATO Air Group 2 initiated the Aircraft, Launcher, and Weapons Interoperability (ALWI) study. The study had the remit to gather data on existing aircraft and weapons, identify where interoperability existed and outline what would be needed in the future to have increased interoperability. The ultimate aim of NATO was to achieve a 'Plug and Play' capability for the systems integration of weapons although the study postulated that for such a capability to be realised, significant technology developments would be required. This chapter will outline the 'Plug and Play' concept and relate the work sponsored by NATO covering ALWI and two subsequent studies relating to the advancement of the 'Plug and Play' concepts and explain how these have paved the way for 'Plug and Play' weapons integration capabilities to be realised over the next few years.

11.2 Systems Integration Considerations

As explained in previous chapters, the level of standardisation achieved to date has centred on electrical interfaces, mechanical interfaces and data formats. However, when integrating a new store, there is the need to modify system software to enable the store to be recognised and

---

Aircraft Systems Integration of Air-Launched Weapons, First Edition. Keith A. Rigby.
© 2013 John Wiley & Sons, Ltd. Published 2013 by John Wiley & Sons, Ltd.

to introduce the required new functionality. Within the United States, a level of commonality has been introduced into the ICDs for several Air-to-Ground weapons such that the data parameters generated by the aircraft subsystems are re-usable across a number of weapons. The interface data space is constructed such that the same data would use the same format and be located in the same messages and sub-addresses of the data bus message structure. However, weapons are allowed to also employ data entities that are specific to their functionality. Whilst these weapon-specific parameters are accommodated in the ICD, it means that full standardisation at the ICD level cannot be achieved. So although standardisation can reduce the magnitude of the systems integration effort, it does not overcome the need for some change to the aircraft subsystems. However, whilst this can have a direct impact on reducing integration risk and timescales (and therefore costs), the level of system change can still be significant.

Europe has so far not adopted this approach largely due to the different nations procuring weapons for their own needs, few new weapon development programmes and there being little drive to have such a level of standardisation across different programmes. This US initiative, whilst enabling existing functionality to be re-used, does little to build synergies between systems and subsystems across multiple platforms. Therefore, whenever a weapon has to be integrated across a number of platforms, there can be little synergy between the specific changes required to the subsystems on each platform. It is therefore not surprising that the total cost of introducing a new weapon into a nation's inventory is proportional to the number of platform types to which it has to be integrated. For example, in the United Kingdom, with the Tornado and Typhoon platforms in service, there is little synergy between the two system modification programmes required to integrate a new weapon. Therefore the total integration cost to introduce a new weapon to the UK front-line fleet will be around twice the cost of integrating the weapon with, say, the Tornado (assuming that the integration costs for both platforms are equal, which in reality may not actually be the case due to different system implementations, international work-share responsibilities, specific technology, etc.).

It is therefore generally accepted that the cost of integrating a new weapon into the inventory of an aircraft and the time it takes are unacceptable and ultimately unaffordable. As discussed in previous chapters, the task of weapons integration involves far more than agreeing the aircraft/store ICD. It includes the introduction of systems modifications, the aeromechanical aspects, safety engineering and so on. From work undertaken separately by NATO and the UK Ministry of Defence, it is not uncommon for the systems integration aspects of weapons integration to amount to 40% or more of total programme costs. Therefore, this is an area that has been identified for further work to simplify the changes required, thereby reducing the timescales and therefore the costs of integration.

Once a weapon has been integrated with the required platforms and introduced into service, its effectiveness will reduce throughout its lifetime. This could be due to a number of factors such as improved target protection measures/countermeasures, the changing face of warfare, new international laws, the need to reduce the cost of supporting the weapon stockpiles and so on. This may lead to the need for the WDO to introduce weapon upgrades which could include new/different warheads and propulsion systems, improved seekers/guidance systems and new fuzing mechanisms. Whilst the WDO would be expected to ensure that the improved weapon is still operable by any aircraft with which the baseline weapon has already been integrated, the full capabilities, particularly for new seekers or guidance capabilities, may not be fully exploitable. To achieve this there may be the need to introduce subsystem updates to the aircraft. Clearly this will incur an expense which will be proportional to the magnitude of the systems and software changes required and also the magnitude of system re-certification effort. Ideally, the aim would always be to minimise such changes. However, at the start of the original programme to develop the aircraft systems, areas which could be affected by the introduction of new/modified weapons throughout the programme's life are not usually considered, and this can result in complex architectural, hardware and software changes being required to integrate a new weapon. Integration costs can therefore escalate especially if any safety-critical functionality is affected, for example, in the SMP. This is a major shortfall that is not addressed by the adoption of weapons interfacing standards and is therefore a significant driver of the through-life costs associated with weapons integration.

The commercial world of information technology has long realised that system and software upgrades are an integral part of cost-effectively introducing new functionality and capability. This is generally known as 'Plug and Play'. We are all familiar with the computers on our desks or in our homes that can be easily upgraded with new hardware and software. But to get to that position, it has been necessary for the commercial computing and software industries to invest billions of dollars and many years to develop and perfect these concepts.

In an ideal weapons integration world, it would be possible for the WDO to develop a new weapon, and this would be delivered with a data file consisting of an appropriate set of information. The weapon would have standard mechanical and electrical interface connections. The aircraft with which the weapon has to be integrated may have no prior knowledge of the new weapon but would provide a standard software interface that is compatible with the weapon data file. The data would be read from the data file and processed within the aircraft's subsystems so that aircraft resources can be allocated to provide the functions that are necessary to operate the weapon. The aircraft subsystem would then re-configure so that the aircraft system as a whole can provide all the correct data entities to the weapon when required. The aircraft system would also obtain knowledge of the weapon launch sequence timelines and safety parameters and would be able to generate the relevant cues to the aircrew. Target data would be loaded in a common format and the aircrew and weapon made aware of the details. On route to the launch point, the aircrew would be given the necessary cues at the right time, and the appropriate data would be transferred to the weapon. Weapon priming would be done automatically (with aircrew intervention only when safety reasons dictate), and the launch point would be calculated using a standard algorithm with the appropriate coefficients having been retrieved from the weapon data file. The weapon would be launched at its optimum release point against its target, optimising its P_k while satisfying aircraft safety requirements. All this without the need to change any system software! If this could be achieved, we would have what is referred to as 'Plug and Play' weapons integration.

11.3 The Journey to 'Plug and Play' Weapons Integration

Clearly, this vision is an idealised vision of the future, and there are many significant technology developments required before this could be realised. Although all the aircraft subsystems could be updated as part of a single programme to support such technology, this could be prohibitively expensive. It is therefore unlikely that this scenario will be introduced to European programmes as a 'big bang' retro-fit as the cost of designing, developing and qualifying such a generic system will be high. It is more realistic to assume that 'Plug and Play' weapons integration will evolve through a series of incremental steps. This is the journey to 'Plug and Play'.

It must be remembered that this type of 'Plug and Play' concept is applicable only to the systems integration activities. Although this should realise significant cost savings, there are many integration activities that will not benefit from this. For example, the aeromechanical aspects of weapons integration will be unaffected. It will still be essential to perform, for example, wind tunnel testing, modelling and test flying to prove that safe separation of the weapon from the aircraft can be achieved throughout the flight envelope. This will have to be repeated for all aircraft types on which the weapon is to be carried. However, as noted earlier, the systems integration activities could form up to 40% of the costs of integrating a new weapon. Therefore, if such technology can generate a significant reduction in these elements of the integration programme (noting that this excludes the costs associated with developing the platform to be compatible with 'Plug and Play' concepts), then this represents a massive potential cost reduction in real terms.

11.4 'Plug and Play' Technologies

The purpose of 'Plug and Play' is to allow weapon system improvements to be implemented incrementally and easily throughout the life of a platform. The term 'Plug and Play', whilst being analogous to that used for commercial Personal Computer systems, would enable new features and functionality to be added quickly, affordably and with minimal disruption to the availability of the existing functionality. Whilst this analogy is useful, it is not fully applicable to weapons/platforms due to the limitations imposed by their existing physical and logical architectures or due to additional requirements to achieve non-functional attributes such as safety or performance. Whilst there is great interest in 'Plug and Play' across a large number of stakeholders (platform operators, platform manufacturers, system integration organisations, weapon manufacturers, airworthiness organisations, etc.), the current lack of a standard set of technologies creates ambiguity across the stakeholder community. This ambiguity has also led to different technologies being developed, each with the aim of reducing the systems integration costs and timescales and providing a weapons integration capability that goes some way to satisfying the goals of the 'Plug and Play' vision. In the remainder of this chapter and the following chapters, some of these key technologies will be reviewed. These include technologies outlined by the NIAG ALWI studies, UAI developed in the United States and open systems enabling technology.

11.5 Adoption of 'Plug and Play' Technology

An important attribute of any 'Plug and Play' technology is that the airborne system must be re-configurable, ideally without the need to change existing certified software.

All computers in the aircraft system that are involved in the integration of a new weapon need to have a method of being re-configured. Ideally, the method of achieving this would be standardised so that each affected computer would have a common method for introducing the new/modified run-time functionality. Such a concept is likely to raise significant integrity issues that need to be addressed in the system design and certification strategy and, as such, could be very difficult to implement in legacy software which was never designed with such capability in mind. Proprietary system architectures across different platforms also militate against standardisation.

Modern avionics computing architectures such as that provided by the Allied Standard Avionics Architecture Council (ASAAC) do provide for new software applications to be easily added in a controlled fashion, thereby potentially easing the introduction of 'Plug and Play' technologies. Even so, to introduce 'Plug and Play' functionality across an airborne system in a single step could be very expensive. However, having a standard 'Plug and Play' method that can be built onto system computers based on a standard architecture could mean that elements of the technology may be introduced progressively. This would mean that as each computer is updated, some of the overall benefits would be realised. If the introduction of the 'Plug and Play' capability was coupled with other upgrades such as those required to overcome obsolescence, then this could be a cost-effective way of adopting such technology. Whilst this approach will be more difficult for older platforms, a lower level of benefit could be seen if only some of the computers where to be updated (e.g. the replacement of an obsolete Mission Computer could benefit from applying 'Plug and Play' technologies). This potential 'Plug and Play' implementation would still need un-receptive computers to be modified in the traditional way when integrating a new weapon, but the new computer would be largely unaffected. Clearly, in such a scenario, overall integration timescales may not be significantly reduced, but costs would be, as there would be one less computer software programme to modify.

To introduce any level of 'Plug and Play' capability will require significant investment in technology and process development. However, investment will only be made if a significant return can be achieved. As such, before any programme is initiated to adopt 'Plug and Play' technology, the business case must be understood and the number of weapons that would need to be integrated to reach the break-even point defined. This is particularly the case for legacy aircraft, where the business case must be based on the number of future stores which will be integrated and offset any potential reduction in integration costs against the investment in technology and process adoption. Clearly, having technology available cannot be the sole justification for its adoption. However, for a new platform design, where it is anticipated that a range of stores will be integrated throughout its lifetime, the business case for adoption should be easier to justify.

As 'Plug and Play' technologies continue to follow a development road map, with both work aimed at delivering near term capability (e.g. UAI) and work aimed at developing the technologies and techniques for greater integration flexibility (e.g. ALWI), the goal of significantly reducing the systems integration costs associated with weapons integration is now closer to being a reality. The remainder of this chapter will give an overview of the NATO ALWI studies, Chapter 12 will discuss the adoption of open systems technologies which can ease the introduction of 'Plug and Play', and Chapter 13 will review the UAI and its NATO equivalent (NATO UAI – NUAI).

11.6 Introduction to Aircraft, Launcher and Weapons Interoperability

In general, the NATO nations procure their own platforms and weapon systems to suit national defence requirements. Whilst standards such as MIL-STD-1760 have provided a level of interoperability, as noted earlier, there has been little effort to standardise the systems implementation. In operational terms, this means that stockpiles of various weapons have to be moved into theatre to support operations, even though some weapons have the same basic capability. Clearly, this is a high logistical overhead.

In 1998, NATO Air Group 2 initiated a study (undertaken by NIAG) to investigate how a greater level of interoperability could be achieved across fighter aircraft, their launchers and Air-to-Air missiles. This initial study paved the way for the development and the thinking around the concepts that would form the basis for 'Plug and Play' systems and more.

This chapter will review the aims and conclusions of the original study and two subsequent studies (ALWI-2 and the ALWI Common Interface (ALWI-CI) studies) and outline the links to many of the subjects relating to standards and systems integration issues covered in previous chapters.

11.7   ALWI Study

The original study undertaken by NIAG was aimed at identifying means to achieve greater interoperability between aircraft, launchers and missiles (hence the term ALWI – Aircraft, Launcher, and Weapons Interoperability). The original NATO view was that interoperability would be achieved when an aircraft could utilise a weapon that was not in its current inventory, on a mission normally carried out with a weapon that was already in the aircraft's inventory. It was envisaged that any changes necessary to the aircraft subsystems to enable the new weapon to be operated would be minimal and require only a short-term action without the need for highly skilled personnel. To this end, the study consisted of three parts. The first was to gather interface information on all relevant aircraft, launchers and Air-to-Air missiles. The second part was to consider a standard set of launcher and missile combinations for each aircraft. Initially, mechanical interoperability was considered and recorded, and this resulted in some combinations being identified as not realistically offering interoperability. The realistic combinations were then analysed in terms of their electrical interfaces. This identified additional combinations which were non-interoperable. Systems implications were then analysed in order to identify the areas that needed to be considered such as the level of hardware modifications required, the level of software modifications required and the level of interoperable performance potentially achievable. Although this did identify a number of opportunities which were reported to NATO, any further action was left to national Governments to investigate.

The third part of the study was probably the most far-reaching, identifying some significant potential solutions for improving interoperability and reducing the costs and timescales of weapons integration programmes. A key short-term recommendation was the need to develop a common ICD to which weapons and launchers would be developed to be compatible with. Whilst the original ALWI study was centred on Air-to-Air missiles, this short-term recommendation was subsequently adopted in the United States as the catalyst for UAI (see Chapter 13). A Common ICD was identified as a means of significantly improving the potential for interoperability in much the same way as the adoption of the AIM-9 Sidewinder analogue interface has become a de-facto standard for short-range Air-to-Air missiles. Also the Common ICD was seen as a key first step towards greater standardisation and the eventual development of 'Plug and Play' concepts, the aim being that if all weapons used the same interface signals, data and functional interface, then any new weapon complying with the interface would be integrated without significant system changes. In reality, it is unlikely that all future weapons could be compliant as each will require its own interface functionality and timelines and possibly weapon-specific data.

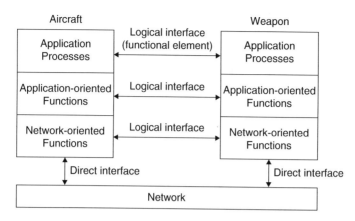

Figure 11.1 Concept of a layered interface framework

The study made two medium-term recommendations. The first was that a layered interface framework should be introduced to the Common ICD. Whilst this recommendation spawned GASIF (defined by AIR5532 – see Chapter 7), it also postulated that such layering of data communication functions would aid in the minimisation of system changes if, for example, a new data communications medium was adopted. The layered interface framework recommendation aimed to separate the functional element of the aircraft/store interface from the application-oriented data protocol that supports it and also to separate the application-oriented data protocol from the supporting data network interface. This would then allow the Common ICD to evolve to accept alternate data network interfaces whilst preserving a common application-oriented data protocol and functional element. More recently, this has been exploited by both AS5725 (Interface Standard, Miniature Mission Store Interface) and AS5726 (Interface Standard, Interface for Micro Munitions), where the standard data protocol available in MIL-STD-1760 has been adopted, but with different physical, data link and network layer interfaces. The concept of a layered interface is depicted in Figure 11.1.

The advantages of adopting such a framework are that it can easily reflect different store carriage configurations, including smart S&RE, and that it enables the management of planned upgrades to existing interfaces.

The second medium-term recommendation of the study was to standardise software functions associated with weapons control such as LAR calculations (a common LAR algorithm), management of S&RE, initialisation of weapon navigation systems and GPS receivers, mission data downloading and so on. Modern systems architectures are based on layered systems architectures which enable software applications (standardised software functions) to be easily integrated (see Chapter 12). Such functions (or services) provide an abstract and standardised view of the weapons and would be developed to be aircraft-independent. The level of abstraction can also be such that common functions can be weapon-independent. Abstraction also has the added benefit that standard software functions can be re-used across multiple systems. This open systems approach to aircraft weapon control software also provides an opportunity to reduce software/system integration costs by isolating the detailed weapon control functions from the mission software and isolating the avionics architecture and weapon station details from the weapon control functions.

The long-term recommendations from the original ALWI study pointed the way to developing a 'Plug and Play' capability, although it was recognised that there would need to be developments in software technologies, particularly in relation to safety-critical software and certification, before this could be realised. The study postulated the development of an aircraft/weapon Application Program Interface (API) and Weapon Data Load Module which would contain data to configure the system. It was postulated that such a Weapon Data Load Module could be either uploaded from the weapon or generated within the Mission Planning System. As we will see in Chapter 13, this concept was also developed further into the UAI implementation.

One final note from the study made reference to the work of the SAE to develop CLARA.

11.8   ALWI-2 Study

Whereby the first ALWI study addressed Air-to-Air missiles, covering all interface elements of interoperability, the ALWI-2 study considered Air-to-Ground weapons, with the main emphasis being on logical (i.e. data/software) interoperability. The ALWI-2 study also considered physical interoperability for both Air-to-Ground and Air-to-Air weapons, making recommendations on how the various mechanical interface standards could be enhanced to foster greater interoperability. The application to both manned and unmanned combat systems, including fixed-wing aircraft and rotary wing aircraft, was considered as were all types of Air-to-Ground weapons, including miniature munitions.

However, the main aims of the study were firstly to develop the original ALWI study short-term recommendation to identify an API and define its requirements with consideration being given to the adoption of Open Systems Architecture methodologies. A second aim was to define a layered interface framework and define standardised weapon control functions. These aims were considered from both the avionics architecture and standards perspectives and also considered early implications from a wider, network-centric viewpoint.

The study identified the need to create a framework that would host 'Plug and Play' store control components within a service-based architecture. This framework aimed to provide independence from the underlying avionics architecture and platform services, whilst exploiting the performance provided by the underlying system. The key to achieving this was to acknowledge that different aircraft could have very different system architectures, meaning that the achieved performance would be unique to the actual implemented system. The 'Plug and Play' store control components were described using a computing platform independent model, with the intention that components (and the model) would be portable from one aircraft system to another. Once integrated in the aircraft subsystems, it was envisaged that the 'Plug and Play' store control components would support the rapid introduction of new stores. With the intention that a new store within a pre-established class and load-out was to be integrated, the aim was to avoid any changes to the embedded aircraft software. The ALWI-2 study envisaged that this would be achieved by loading the parameters for the new store into the aircraft system at store discovery time.

Whilst this would enable a new store that was compatible with the existing capabilities of the aircraft system to be integrated without changes to the system, a new store of an unknown class would not be compatible. In this case, the ALWI-2 study identified that a new store control component for that class would have to be integrated. The benefit of having standard modules complying with a standardised API would mean that the module code and computing platform independent model may already have been developed for another aircraft. In such a case, the systems integration task would be reduced to ensuring that the module could be executed in the presence of existing software on the underlying computing infrastructure. This approach also would deliver an additional benefit that once the new store class had been integrated, subsequent stores within that class could be integrated without further change to the aircraft's software.

In order to understand how such API functionality could be realised, the ALWI-2 study recognised that the real aircraft computing system would need to be described using a common conceptual framework which would set the vocabulary and provide a common set of Technical Reference Models (TRMs) for describing standardised interfaces on which weapons interoperability and therefore 'Plug and Play' depends.

Each TRM identified by the ALWI-2 study provides a different and independent view of open systems and allows coordination between different standardisation activities within a common Technical Architecture. For the aircraft avionics architecture, two TRMs were identified and profiled to meet the needs of aircraft/store interoperability. These are GASIF, as defined by AIR5532, and the Generic Open Architecture (GOA) Framework defined by AS4893. The distinction between the two frameworks is important. As described in Chapter 7, GASIF is an appropriate TRM for describing aircraft/store communication protocols such as MIL-STD-1760. The GOA Framework applies to the development of Open System Architectures, where the layers and interfaces are 'real' and must be implemented by computing platform components. As the internal architecture of weapons is likely to be very different for different weapons and that the aim of the ALWI-2 study was centred on the aircraft, the GOA profile was restricted to defining an abstraction of the aircraft system (see Figure 11.2).

In the TRM, the computing infrastructure of the aircraft (denoted Application Platform in the figure) consists of the Physical Resources (processor, memory, input/output devices,

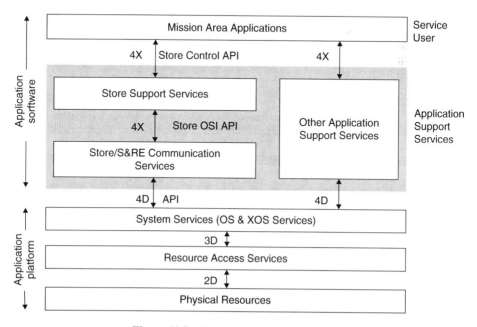

Figure 11.2   GOA profile of aircraft system

bus transceivers, etc.), Resource Access Services (software device drivers) and System Services (the basic software Operating System (OS) providing an API on which software applications can be interfaced) and Extended OS (XOS) services which provide extended capabilities within the system (such as Blueprint management – see Chapter 12)). The GOA framework identifies a number of direct interfaces (2D, 3D, etc. in the figure – it should be noted that a 1D interface also exists in GOA, but this would be a connection to an external data transfer media such as a data bus and, as such, is covered by the GASIF TRM) which provide the lexicon for describing system components and where they fit within the overall system architecture.

The Application Platform provides a 4D API on which Application Support Services (part of the overall system Application software) can be integrated. It is this part of the system architecture where standard services exist. These could be, for example, services which provide data communications with stores, control of S&RE, inventory management and so on. It is also these services which enable a level of system standardisation and software component re-use to be realised. This is a basic requirement for 'Plug and Play' functionality.

Where individual services need to communicate with other services in the Application software, the GOA profile allows this (identified as 4X interfaces). Within the Application Support Services, the Store OSI API provides software access to the interconnection environment of the aircraft/store interface. This is fundamental to aircraft/store 'Plug and Play' as it facilitates cross-platform portability of the Store Support Services Software.

Finally, where a store requires specific functionality that cannot be provided by common standardised services, then specific software applications can be implemented (Mission Area Applications in Figure 11.2). The Application Support Services provide a Store Control API to these higher level weapon control applications.

The TRM is defined to be compatible with existing Open System Architectures such as the ASAAC architecture (see Chapter 12).

Most weapons, when their functionality is abstracted, will employ a common set of generic functions such as initialisation, system testing, Transfer Alignment, mission data download-ing, weapon fire and so on. Within the Application Support Services, a number of abstract services can therefore be defined. Typical services are likely to include Store Support Services, Store Manager Support Services, Store/S&RE Interface Services and Store Communication Services. Table 11.1 outlines some of the typical service types and what they provide for the weapon control system integrator.

Undertaking an analysis of the operations required for various stores (ballistic bombs, smart munitions, carriage systems and various types of pods, covering both Air-to-Ground and Air-to-Air) identifies a large number that are used to control stores. These operations can then be categorised into functions (i.e. a single discrete event such as application of the MIL-STD-1760 Release Consent signal) and services (i.e. a series of events which would normally be con-trolled by a state machine, e.g., Transfer Alignment). These operations can then be mapped to either the Store Control API or Store OSI API identified in the TRM.

Such a categorisation enables a level of standardisation to be adopted. Therefore, whilst the TRM provides an abstract model for understanding the basis of a 'Plug and Play' capable aircraft weapon control system, it is important to also produce re-usable, standardised 'Plug and Play' components. For this, the components would be modelled using Object Oriented principles, for example, using the Unified Modelling Language (UML). By modelling functions at an abstract level, the actual behaviour of the implemented functionality can then

Table 11.1   Typical service types

Service name	Service function
Store Support Services	These services contain the store classes that provide the services to the Mission Area Applications as identified in the Store Control API. These include functions to power, select, arm, initialise and release the stores carried on the aircraft.
Store Manager Support Services	These services provide all of the facilities required by the Stores Manager that are not directly related to an individual store and therefore are not provided as part of a store class. This includes aircraft functions associated with store control, for example, BIT results storage, Elapsed Time Indicator, Weight On Wheels and so on. These services are expected to be rarely modified during the life of the platform.
Store/S&RE Interface Services	These services implement the services provided by the Store OSI API that control the interface with the S&RE.
Store Communication Services	These services implement the Application, Presentation, Session, Transport, and parts of the Network layers of the OSI communication stack (see Chapter 7). These layers enable data communication interfacing with the underlying store interface (e.g. MIL-STD-1760).

be defined by data contained in a configuration file. This is the basic concept which underpins 'Plug and Play' technology.

A key element of a successful 'Plug and Play' capability will be the development and documenting of the Store Control API. To be ultimately successful and to foster interoperability, the API must be defined as an open standard.

Once the basic concept of standardised functions and services (implemented as object-oriented software applications) is adopted, then the concept can be extended to provide broader aircraft system platform 'effects' services to the network-enabled battlespace.

The ALWI-2 study also reviewed the multitude of standards which could be adopted to enhance aircraft/store interoperability. These included the weapons integration-specific standards discussed in earlier chapters, but also included a number of other standards such as those covering certification, commercial data buses, system architectures such as ARINC 653 and ASAAC, and environmental standards. The study recommended that an overall standards Technical Architecture should be evolved to support 'Plug and Play' goals. The aim of this recommendation was to foster the greater standardisation of a smaller number of standards, some of which would need to be written/updated to focus specifically on supporting a 'Plug and Play' capability. To ultimately achieve an interoperable stores 'Plug and Play' capability, a consistent, non-conflicting set of non-proprietary standards will be required.

11.9   ALWI Common Interface Study

The first two ALWI studies had identified the initial concept of 'Plug and Play'. The ALWI-CI study developed the concept to a greater level of definition by addressing a number of aspects of systems integration. These included an investigation of current and evolving standards and requirements for new standards that would need to be developed to increase interoperability. The study also considered a level of 'Plug and Play' based on a common ICD concept and a more advanced form based on a common Store Control Service (including store discovery), available to the platform or to remote service users. Finally, it was recognised that for 'Plug and Play' concepts to be successful, implementation issues such as certification, life-cycle cost implications, the ability to integrate legacy weapons and compatibility with standardised implementation architectures would need to be considered.

The study group's response to each of these aims will be discussed in the following sections.

The ALWI-CI study continued to develop the thinking around the services provided by an aircraft/weapon combination to a wider, network-enabled battlespace. In doing so, the study developed a number of views (Operational Views, Systems Views and Technical Standards Views covering ALWI interoperability concepts) in accordance with the US Department of Defense Architecture Framework (DoDAF).

11.9.1   Technical Architecture

The Technical Architecture postulated during the ALWI-2 study was further developed to define an organised set of existing and emerging interoperability standards and identified additional standards that would be needed to complete a formal ALWI Technical Architecture. The purpose of this work was to foster a coordinated approach to standards development. A fifteen years timeframe was considered (starting from 2004) in the Technical Architecture with the aim of it being the principal planning document for weapons integration standards development and adoption. A high-level view covering the key areas for standardisation is shown in Figure 11.3.

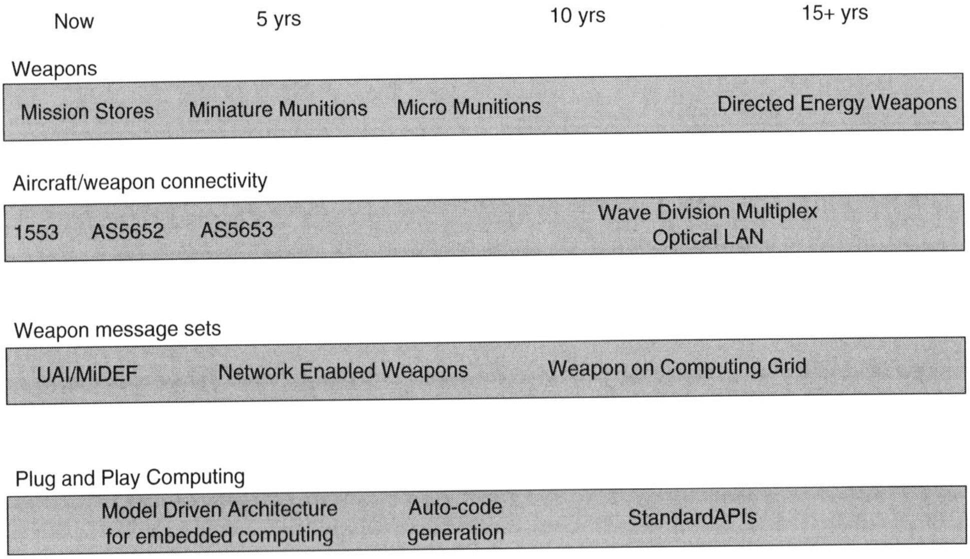

Figure 11.3   ALWI-CI Technical Architecture areas for standardisation

11.9.2   Greater Interoperability through a Common ICD Approach

The objective of adopting a common ICD approach is to standardise the interface between stores and aircraft and to remove the need for aircraft to support multiple unique interfaces. This concept was outlined by the original ALWI study, the aim being that aircraft which have implemented the standardised common interface would have the capability to integrate all stores which comply with the interface without having to change system software. This has a potential direct impact in substantially reducing both integration costs and the amount of time required to integrate a new store. As a new store would be integrated without the need to change system software, it would be possible to introduce new capability to the operational air force fleet outside the usual periodic upgrade cycles.

11.9.3   Common Store Control Service

An extension of the common ICD approach to fostering a greater level of interoperability can be achieved through a series of common ICDs and configuration data files. The approach is based on a set of common interface control documents, tools and processes that address weapon integration on aircraft and Mission Planning Systems to support all phases of weapon employment. This approach forms the basis of the UAI initiative (see Chapter 13), and the study recommended that this should be extended into a NATO-wide implementation. Whilst UAI provides a useful means of introducing new weapons capability to receptive aircraft systems, the ALWI-CI study also developed the true 'Plug and Play' concept.

The study further developed the methodology for partitioning the aircraft software into hierarchical service domains that could be accessed via standardised APIs. To achieve platform-independence (independence from the aircraft systems' computing platforms), the study recommended that these services be defined according to the Object Management Group's Model Driven Architecture® (MDA®) approach using the Executable Unified Modelling Language (xUML). Possible approaches to generating aircraft-specific system designs and software components from xUML models (whilst noting that each aircraft programme may adopt its own approach without affecting interoperability) were defined by the ALWI-CI study. By providing such a methodology, the 'Plug and Play' concept is abstracted away from the actual implementation in the real system. This level of abstraction is a key enabler, not only for providing a 'Plug and Play' capability across multiple system architectures, but also as a means to avoid the effects of hardware obsolescence.

The MDA® approach defined by the study is based on integrating the ALWI service domains with other programme-specific domain models to create a platform-independent model of the aircraft system. This can then be translated into a platform-specific model for the system implementation. The study investigated how MDA®-based specifications can be realised in aircraft hardware and software, giving special consideration to the implementation of an aircraft system based on the ASAAC architecture. The MDA® approach is defined in more detail in Raistrick et al. (2004).

Typical interoperable software components that are central to the 'Plug and Play' concept are shown in Figure 11.4 as aircraft class packages, or domains.

The Weapon Control Domain provides a common Store Control Service that interrogates all store stations on the aircraft. When the stores inventory is being initialised, as each store is

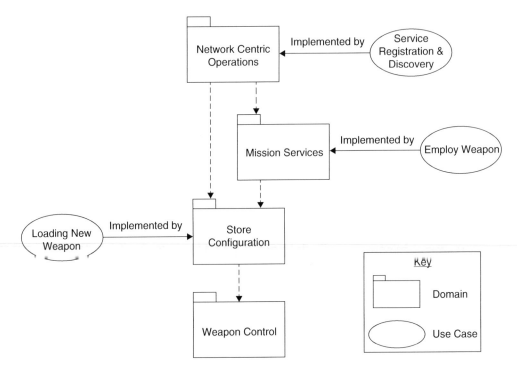

Figure 11.4 'Plug and Play' domains

'discovered', one instantiation of the actual weapon class will be created. The domain will provide all of the necessary software classes to control a single weapon at a station. In doing so, a single domain model is capable of handling all stores including all UAI-compliant MIL-STD-1760 weapons. By abstracting functionality into higher level domains, the same domain model can be shared by all aircraft, irrespective of their computing architecture. Access to the domain may be provided by a standard API.

In Figure 11.4, the client for the services provided by the Weapon Control Domain is the Store Configuration Domain. This domain provides store status and control services to other client domains while satisfying the flight and safety requirements of the aircraft (e.g. maintaining store balance constraints). The domain implements the Loading New Weapon use case that requires the characteristics of each weapon to be described in a set of data files (the store data), with the aircraft characteristics associated with the complete weapon load-out being defined by a different set of data files (load-out configuration data).

Two domains are possible clients for the Store Configuration Domain. These are the Mission Services Domain and the Network Centric Operations Domain. The Mission Services Domain provides various services associated with the Employ Weapon use case shown in the figure. Services may include association of weapon type with target, mission planning for weapon type (including accepting or generating MiDEF data files) and fire control. The Network Centric Operations Domain will provide the structure services for aircraft that have Network-Enabled Capability.

11.9.4   Model-Driven Architecture Approach

As noted earlier, the 'Plug and Play' concept defined by the ALWI-CI study is based on the MDA® approach. MDA® is a framework for software development based on models of varying levels of abstraction from the underlying system architecture. Three types of model are central to the approach. These are the Platform-Independent Model (PIM), the Platform-Specific Model (PSM) and the Platform-Specific Implementation (PSI).

A PIM describes a software system at an abstract level that is independent of the aircraft's computing hardware, operating system, software infrastructure and programming language (in this instance, 'platform' relates to the computing infrastructure rather than the aircraft). By abstracting functionality away from the actual system implementation, common components within the PIM may be re-used as the common functionality model across multiple system implementations (i.e. it can be re-used across different aircraft). The Common Store Control Service is defined as a set of modular components within the PIM. The PIM, defining elements of the overall aircraft system involved in the control of weapons, is likely to also include components that are not standardised, for example, the software components associated with aircrew controls and displays. The 'Plug and Play' capabilities delivered by use of the MDA® approach involve building an architecture made up of Data-Driven Domains where each Domain addresses a single subject (e.g. store configuration). The methodology enables subject matter experts to develop abstract, long-life models independent of the system implementation, which can be automatically translated into software. This results in components that are highly re-usable, enabling new or modified functionality to be introduced in a modular fashion. Changes in the model can be incorporated with relative ease, thereby reducing cost and timescales. It is also possible to conduct operational simulations of the capability at the PIM stage to support requirements development.

Once the PIM has been defined, it can be transformed into one or more PSMs. It is this stage where the aircraft computing hardware, operating system, software infrastructure and programming language are defined. This transformation enables the PIM to be mapped to the underlying system architecture. Therefore, the PSM will differ between the system architectures of different aircraft. The mapping from PIM to PSM lends itself to the application of automated tools. This means that there are potentially great efficiency savings to be made in developing the system implementation.

The final step in the software development life cycle is the transformation of each PSM into a PSI, or source code. Again, the use of automated tools will significantly improve software code productivity.

Figure 11.5 shows a simplified view of the aircraft system software development life cycle based on MDA®. While a single PSM stage is shown, the PIM to PSI transformation may occur through any number of PSM stages ranging from none (in a very simple system, a direct mapping of the PIM to the PSI) to many PSMs that reflect layered choices about the targeted avionics architecture such as programming language, infrastructure services, operating system and hardware.

In the MDA® process, the transformation from PIM through PSM to PSI would be defined for the target aircraft systems architecture. Once defined, new service components providing new functionality can then be added to the PIM and quickly transformed into source code. This is a much more efficient process than updating a paper specification of the system requirements and changing the source code manually for each aircraft implementation.

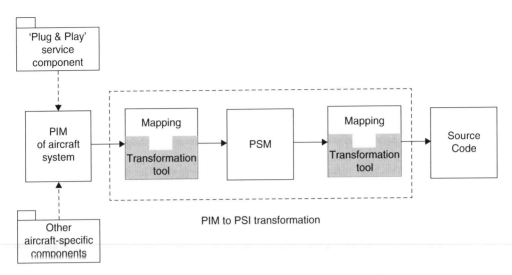

Figure 11.5   'Plug and Play' PIM to PSM transformation

The MDA® approach offers significant benefits to the aircraft system designer. The ability to configure the functionality of the PIM by changing configuration data enables aspects of system behaviour that are subject to frequent changes (e.g. weapon release sequences) to be defined by data tables rather than by changing software code. The extensibility of the approach also enables the introduction of new functionality without impinging on existing certified software components.

The obvious advantage of the MDA® approach over the traditional software development life cycle is the speed at which new operational capabilities can be introduced to the aircraft systems once the MDA® tool environment has been established. From real systems projects which have adopted the MDA® approach from the outset, it is suggested that software maintenance costs can be reduced by an average of 60%.

The complete MDA® methodology is shown in Figure 11.6. An important discriminator of this approach is that different specialists build different parts of the overall modelling and system implementation. For example, the PIM, which captures knowledge about how weapons work, system operation and so on would capture the relevant domain knowledge of such specialists. The mapping from the PIM to the PSM is largely an exercise to partition the PIM to the underlying system hardware and software architectures. It is unlikely that the Weapons Integration specialist will be expert in this field, so this is undertaken by other relevant specialists. Similarly, the Architecture experts would not necessarily be expert in software coding and testing, and so the PSM to PSI mapping is again undertaken by specialists with the relevant skills.

Once the various mappings have been defined, automated processes can generate the actual artefacts of each stage (PSM from PIM and PSI from PSM). Also, by having these distinct stages, it is possible to run simulations at each stage in order to prove that the artefact at that level is correct and satisfies the requirements. For example, the functionality defined in the PIM can actually be executed in the required time slots provided by the underlying hardware architecture.

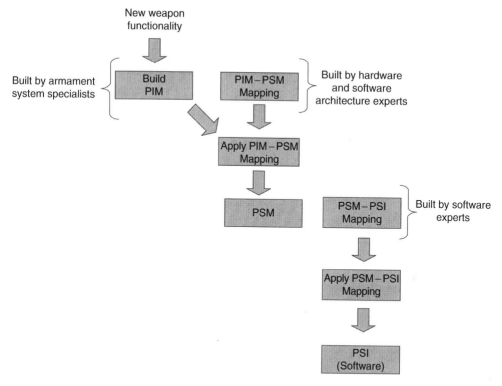

Figure 11.6   PIM to PSI transformation

Once the various models and mappings exist, then the introduction of new functionality (e.g. the introduction of a new of modified weapon) will feed into the PIM. All other transformations are then automatically generated using the existing mappings. It is this level of automation, coupled with the application of the appropriate specialist knowledge in each phase, which provides the significant cost and timescale benefits claimed for the MDA® approach.

11.9.5   Implementation Considerations

A significant benefit of having an abstract PIM means that the functionality can be partitioned as a part of the transformation into the PSM to the underlying system regardless of its architecture. The PSI is the representation of the PSM in the target system elements (software code and computing hardware). When developing the system implementation, there is a need to analyse other aspects that will not have been considered in the development of the PIM. By definition, the PIM has no knowledge of the underlying system implementation and so cannot account for real implementation issues such as processor timing, memory partitioning and computing resource utilisation.

Table 11.2 identifies various aspects of implementation that need to be considered when mapping the PIM to the real system.

Table 11.2 Implementation considerations

Implementation considerations	Comment
Safety Analysis	How will the implementation architecture address the need for fault isolation and avoid the propagation of the effects of faults in the wider system? How will interference be avoided across a safety partition in memory? What is the desired level of reliability, availability and maintainability?
Resource Utilisation Analysis	Will the additional processing demands overload the processors? What is the required data precision/accuracy? Will the additional data fit into the available memory? How will data integrity be assured? Will the data transactions overload the data buses or over-run time slots?
Timing Analysis	Which is the most appropriate scheduling approach? Is the system schedulable? Will there be frame overruns and how are these to be overcome? Will the desired response times be achieved? Will the real-time performance be adequate?
Reliability Analysis	Is there a need for cold/warm/hot standby components? Is there a risk of dead-lock/live-lock?
Security Analysis	Are security boundaries violated? What are the needs for data integrity? What are the needs to ensure data confidentiality?
End-to-End Data Flow Analysis	What level of data latency is acceptable? What level of data jitter is acceptable? Will data integrity requirements require error correction? What is the required data precision/accuracy?
Change Impact Analysis	How will change be controlled such that it does not impact the system? How will changes to the system impact system validation? How will the effect of change on areas of the system not directly contributing to the new functionality be controlled?

In analysing these areas, the system designer may need to trade aspects of the implementation in order to arrive at an optimum solution. For example, a redundant implementation architecture using two of more channels introduced to improve safety or availability will increase failure rates as there are more components in the system which can fail. However, if correctly designed, the redundant architecture would contribute to a greater level of integrity and/or availability but with the penalty on the need for more frequent post-flight maintenance.

11.10 ALWI Conclusions

The ALWI-CI study was the culmination of many years of work on Aircraft, Launcher and Weapon Interoperability. The ALWI studies were the fundamental vehicle for defining the concepts that have since been developed into system capabilities that are being adopted by modern aircraft programmes. Many of the recommendations from the three studies have been adopted by several nations and NATO as the basis for improving aircraft/store interoperability, fostering the further development of interoperable standards that support the systems integration of aircraft and stores and developed the thinking that has been central to the development of 'Plug and Play' weapons integration technologies.

The interoperability concept presented by the ALWI studies also identified the route to a greater level of network-enabled capability for weapons, providing a service to the wider battlespace.

Further Reading

Raistrick, C., Wilkie, I., Carter, C., *et al.* (2004) *Model Driven Architecture with Executable UML*, Cambridge University Press, Cambridge.

Society of Automotive Engineers. (1996) The Generic Open Architecture. *Aerospace Standard 4893*, Society of Automotive Engineers, Warrendale.

Society of Automotive Engineers. (2003) The Generic Aircraft – Store Interface Framework. *Aerospace Information Report 5532*, Society of Automotive Engineers, Warrendale.

Society of Automotive Engineers. (2008) Interface Standard, Miniature Mission Store Interface. *Aerospace Standard AS5725*, Society of Automotive Engineers, Warrendale.

Society of Automotive Engineers. (2009) Interface Standard, Interface for Micro Munitions. *Aerospace Standard AS5726*, Society of Automotive Engineers, Warrendale.

United States Department of Defense. (2007) Aircraft/Store Electrical Interconnection System, *Military Standard 1760*. US Department of Defense, Philadelphia.

12

Open Systems

12.1 Chapter Summary

It is envisaged that open system architectures will provide the basis on which a 'Plug and Play' weapons integration capability will be realised. This chapter will discuss how open architecture mission systems coupled with modular weapon function software will have a significant bearing on the future of weapons integration. The chapter will also provide an overview of the ASAAC architecture and will consider the evolution of the contracting and industry environment and how this will pave the way for delivering open systems technology that supports the introduction of a 'Plug and Play' weapons integration capability, thereby reducing integration costs and timescales.

12.2 Introduction

Chapter 4 covered the design of the aircraft SMS. The SMS architectures discussed captured all weapon control functionality within the subsystem. The consequence of this is that both safety-critical and mission-critical functionality are implemented within the subsystem. Although it is possible to partition functionality of different integrity levels within a subsystem, any change to its software, for example, when a new weapon is integrated, would incur a high re-certification overhead.

An alternative SMS architecture could be realised by a relatively simple but high-integrity 'switch' that provides the routing and control of weapons interface signals under the control of modular software applications residing in an open architecture mission system. The aim would be to design and certify the 'switch' (a safety-critical interface controller) once so that it provides the appropriate capacity for all current and envisaged weapons, thereby minimising

Aircraft Systems Integration of Air-Launched Weapons, First Edition. Keith A. Rigby.
© 2013 John Wiley & Sons, Ltd. Published 2013 by John Wiley & Sons, Ltd.

the need for through-life modifications. When new weapons are integrated, then it would be necessary only to add (or modify existing) weapon function software applications residing in the mission system. This shift in the implementation of an SMS towards implementing weapon control functionality in an open architecture mission system, coupled with modular weapon function software, could have a significant bearing on the future of weapons integration.

12.3   The Contracting and Industry Environment

Before explaining how this new approach will provide significant benefits to integration programmes, it is necessary to understand the contracting and industry environment for current system upgrade programmes.

Figure 12.1 depicts the environment and considers four viewpoints: those of the Contracting Agency, equipment suppliers, component suppliers and the ADO. Each of these viewpoints is considered later.

With shrinking defence budgets, it is now more important than ever for the Contracting Agency to improve the military capabilities at lower costs in terms of upgrades and on-going support. It is rare for sufficient funding to be available to perform a number of large capability improvements simultaneously. For this reason, the Contracting Agency will seek incremental, but regular upgrades that give a continuing increase in capability whilst spreading the costs in a more manageable way. This approach demands that industry provide continuous capability upgrades quicker, but with the benefit of potential long-term cash flow into the business. As a trade-off for this benefit, industry is expected to carry more risk. Another aim of the Contracting Agency is to provide the end user with optimised capabilities across the fleet, thereby avoiding 'fleets within fleets' and enabling other support cost reductions to be realised.

Equipment suppliers invest heavily in their bespoke system designs and therefore seek to exploit their designs across a number of platforms. They must have a broad customer base for their technology to be viable and therefore are attracted by large markets (e.g. the United States). The equipment supplier would ideally have their technology optimised across multiple platforms across multiple customers so that economies of scale could be realised.

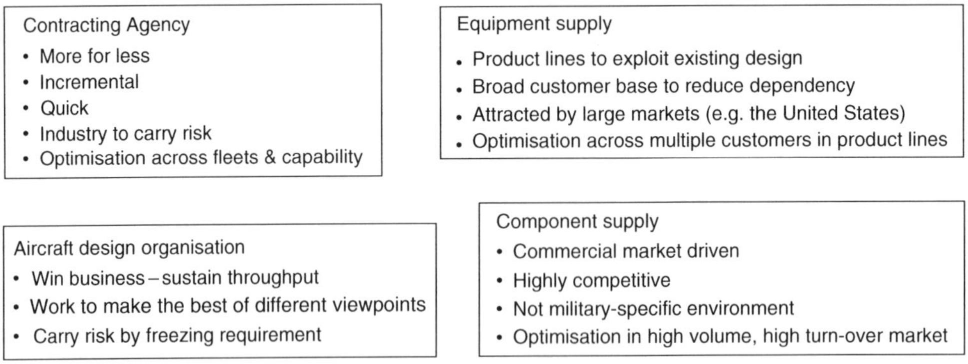

Contracting Agency	Equipment supply
• More for less • Incremental • Quick • Industry to carry risk • Optimisation across fleets & capability	• Product lines to exploit existing design • Broad customer base to reduce dependency • Attracted by large markets (e.g. the United States) • Optimisation across multiple customers in product lines

Aircraft design organisation	Component supply
• Win business – sustain throughput • Work to make the best of different viewpoints • Carry risk by freezing requirement	• Commercial market driven • Highly competitive • Not military-specific environment • Optimisation in high volume, high turn-over market

Figure 12.1   The contracting and industry environment

The supply of electronic components is driven by the needs of the commercial market, particularly industries such as Information Technology and telecommunications. The component supply market is highly competitive, and it is not uncommon for a lifetime buy of a particular component for a military programme to be significantly less than one month's production output of a single manufacturer. For these reasons, the component supply business has little interest in the manufacture of components that are optimised for the military environment.

The aim of the ADO is to balance the demands of the industrial environment and to continue to win new business and sustain throughput. However, the ADO is expected to carry significant risk. To manage the risk, the aim is to freeze requirements early and then to be averse to subsequent change. This can be directly at odds with the needs of the Contracting Agency and is therefore a failing method in the increasingly complex world of weapons integration. However, the ADO sees real business benefits in the incremental capability growth required by the Contracting Agency. For this reason, concepts which encompass 'design for upgrade' are becoming increasingly necessary in order to enable frequent change to be embraced.

Clearly, the environment in which the ADO exists is volatile. With the Contracting Agency demanding more capability in shorter timescales and with technology obsolescence a real in-service issue, the fixed development life cycles cause integration costs to spiral, ultimately leading to upgrades becoming unaffordable. If military customers are to continue to obtain capability improvements in the form of more capable weapons and the ADO and equipment suppliers to have viable businesses, then a new technology approach is required. There is a need for industry to embrace the demand for continual change and upgrade to systems. A first step is for systems to be designed such that a level of application modularity is achieved.

12.4   Current Systems

In current systems, hardware-related functionality and mission-specific functionality are embedded within the operational software. When a new weapon type is added to the inventory, then it is almost impossible to re-use any of the existing software without some modification. Also, as software is modified, the amount of testing required to prove that existing functions have not been adversely affected becomes time-consuming. As discussed in Chapter 11, the adoptions of model-driven design, where commonly used functions (e.g. stores inventory management) are separated out into service domains, will reduce the software development activities and, therefore, will also reduce costs and implementation timescales. However, modularising applications without addressing the potential to manage obsolescence and allow major technical upgrades without disruption will not reap all the potential benefits.

A typical Mission Computer may be based on a bespoke backplane that forces the adoption of other bespoke hardware such as graphics and processor cards. Upgrades due to obsolescence or the need to improve performance require new hardware modules to be developed and qualified.

The use of non-standard hardware means that all software is tailored specifically to each electronic module leading to close coupling between the hardware and the software. If an existing mission system implementation is to be expanded to introduce, for example, additional processing capability (possibly adopting a different processor technology) or new data communications technology, then the existing equipment software will be subjected to major re-work. Existing applications cannot be readily re-distributed onto different processors, and the addition of new, low-level hardware drivers may also require the higher-level functionality software to be changed. This makes the software development activities difficult to partition across work teams leading to large, complex development teams. Also, and very importantly, these constraints generate an inability to mix applications of different safety integrity levels, leading to the whole operational software program being developed to the same integrity level demanded by the function with the highest required level of integrity.

All these changes to the mission system have major impacts on software design, coding, testing, integration and qualification and ultimately cost, risk and timescales.

A level of modularisation of functions in existing systems could be beneficial although it is unlikely that a cost-benefit analysis would generate favourable results unless a mechanism for upgrading the underlying hardware is also adopted.

12.5   A Typical Mission Systems Upgrade Programme

Figure 12.2 identifies the principal activities that are needed to modify and prove a platform's mission system when introducing new capabilities such as integrating a new weapon application. The figure assumes that the weapon to be integrated is already developed and

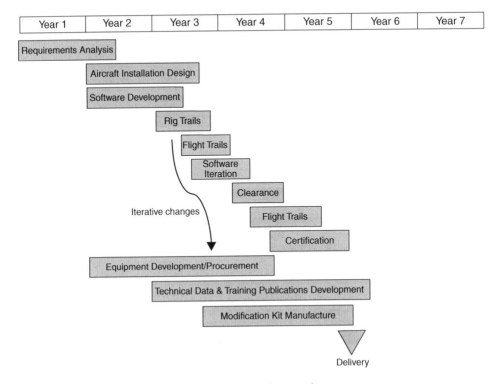

Figure 12.2   A typical avionic upgrade programme

cleared for use on other platforms. In this example, the systems integration programme typically could take up to five years.

The process follows a typical waterfall programme that is largely paced by the development of new software, changes that require the procurement, development and iteration of new hardware, iteration of software to correct errors and sub-optimal features of the initial software load and the integration of these elements into the system. Ultimately, this leads to certification of the upgraded system and delivery to service of the modification package and associated technical data and training publications. Clearly, the length of the programme is a significant cost driver.

12.6   ASAAC Architecture

The ASAAC was formed to develop a Modular Avionics Architecture (MAA) for airborne avionic systems. The ASAAC documentation set defines an open architecture which comprises a distributed software layered architecture, Common Functional Modules (the hardware), a unified communication network (and protocols) and a system management hierarchy. Of these elements, the most beneficial to the systems integration of weapons is the distributed software layered architecture.

Figure 12.3 outlines the open systems software architecture, often referred to as the 'three-layer stack', consisting of the Module Support Layer (MSL), the Operating System Layer (OSL), the Application Layer (AL) and the interfaces between the layers (Module to Operating System (MOS) and Application to Operating System (APOS) interfaces). The distinguishing characteristic of the MAA software architecture is the system management hierarchy, which comprises the Application System Manager (in the AL) and the Generic System Manager

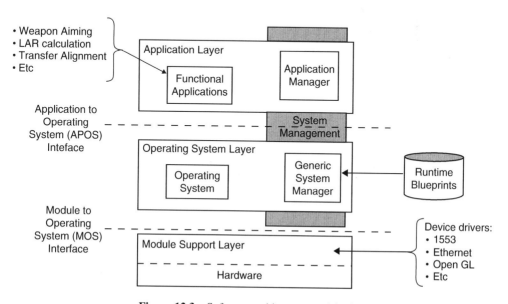

Figure 12.3   Software architecture partitioning

(GSM – in the OSL). The GSM configures the avionics run-time system according to Run-time Blueprints.

By employing standard software interfaces, the effect of change can be controlled such that only those parts of the software architecture which need to be modified (e.g. to add a new device driver) are changed, leaving all other software unaffected.

This open architecture approach enables the migration of new hardware into the equipment to overcome obsolescence issues (e.g. enabling the introduction of a new processor, possibly from a different family (e.g. Power PC®, Pentium®, etc.)) or to introduce a new data interface (e.g. EBR-1553 in place of MIL-STD-1553B), without the need to also undertake major re-design of the software. This makes the adoption of multi-source Commercial-Off-The-Shelf (COTS) hardware products viable.

To achieve a level of hardware independence obviously requires, in the first instance, the equipment's software to be restructured. This restructuring enables the actual hardware to be hidden behind the standard software interfaces. This has a major benefit of protecting the investment made in mission software. As new requirements identify the need for different input/output interfaces or greater processing power, the software is able to be easily re-configured without the wholesale need to scrap the previous version of mission software. An open architecture system is therefore a necessary component needed for change to be embraced.

The MOS interface provides a set of standard device driver interfaces. These would include, for example, input/output devices for MIL-STD-1553B data buses, Ethernet connections or Open GL graphics generation for cockpit displays. The MSL provides services that are supported by the underlying hardware to the OSL. A significant benefit of this approach is that should, say, the MIL-STD-1553B data bus connection need to be replaced, say by a Fibre Channel interface, then the underlying hardware would be upgraded, and only the MSL software would need to be altered in order to drive the Fibre Channel interface. The higher layers would not be affected.

The OSL can be considered in two generic parts. The Operating System software part translates the services provided by the MOS to the higher AL. However, this software needs to be 'managed' so that hard real-time constraints and the required Quality of Service (QoS) that is required by the higher Application layers can be maintained. This is performed by the GSM part of the OSL. It is the responsibility of this block to manage the use of the resources available at the lowest layers in the stack so that real-time and QoS performance can be maintained.

Clearly, with such an open system design, not all the resources/services provided below the MOS may be needed. The OSL can then be configured to use only those resources/services that are required by the higher AL. To do this, a system 'Blueprint' is defined at design time. This is a relatively simple file that is interrogated by the GSM block in order to configure the services that need to be available at the APOS interface and to effectively use the resources/services below the MOS in order to maintain the real-time/QoS demands of the AL.

The AL consists of a number of software objects (applications) that provide the functions of the Mission System. This would include, for example, objects implementing weapon-aiming calculations, LAR calculations or Transfer Alignment routines. Object-oriented design techniques mean that functions can be partitioned so that new functions (e.g. aiming algorithms for a new weapon type) can be easily added without affecting already available software applications for existing functions. To hold this together and to provide the necessary scheduling, the AL also contains the Application Manager function.

There are a significant number of important benefits provided by this architecture. These are:

- As all operating system/application software is independent of hardware, obsolescence management and technology insertion are eased. Hardware changes cause associated changes to the MSL only without impacting on the higher software layers.
- The ability to utilise accepted open commercial hardware standards, for example, Open GL for the generation of displays, means that a common software load between the operational system and training aids is possible, thereby enabling, for example, crew display formats for new weapons to be generated and assessed on a desktop PC and then easily integrated into the real system.
- Applications can be written in any accepted High-Order Language such as Ada, 'C' and so on.
- Mixed criticality applications can be hosted on the same hardware.
- Predictable operation can be maintained. This is particularly important for safety-critical functions such as weapon deployment.

In order to implement an MAA and to be independent of the underlying hardware will require investment in re-engineering existing Mission Systems. This could be costly, and initially, no additional capability may be available. However, the benefits will start to be realised on subsequent upgrades either to introduce new capability (such as a new weapon) or to upgrade obsolete or low performance hardware.

12.7   ASAAC and 'Plug and Play'

In addition to the architecture characteristics described earlier, there are three important features of the software architecture that assure technology independence and support a means of implementation of a 'Plug and Play' concept for stores integration. These are the layered structure, the implementation of virtual channels within the software architecture, and the use of Blueprints.

The embedded computing architecture framework described in Chapter 11 decouples the application software from system services software, and system services software from the hardware. This is totally consistent with the ASAAC MMA shown in Figure 12.3. From Chapter 11, each functional Application required to operate the store is realised as a transformation of the PIM into executable code. The MSL provides the software interface between the operational software and the system hardware, including the aircraft/store physical layer interfaces. The System Manager function determines the allocation of hardware resources (store physical interfaces) to the functional Applications using Run-time Blueprints.

The layered structure of the architecture allows the functional Applications (consisting of one or more software Processes) to be executed seamlessly in a systems architecture that is distributed across a number of microprocessors (e.g. different microprocessors in different computers such as between a Weapon Aiming computer and an SMS). Data is exchanged between peer Applications (logical interfaces) via virtual communication channels, which decouple the Application functions from the physical paths (direct interfaces) implemented by data bus protocols and hardware. Figure 12.4 shows the Virtual Channels (VCs) in the OSL mapped to the Transfer Connections (TC) in the MSL, which map to Network Interface Units

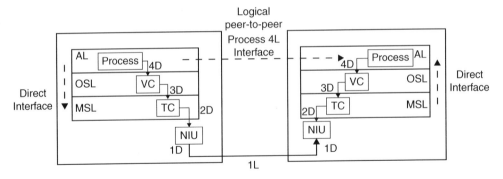

Figure 12.4 ASAAC communication channels

(NIU – e.g. MIL-STD-1553B data bus transceiver) in the hardware. The figure also identifies the associated logical (1L and 4L) and direct (1D, 2D, 3D and 4D) GOA interfaces (as outlined in chapter 11).

For one functional application to logically communicate with a peer application on a separate processor, communications have to be through direct interfaces through the layers (as shown in the figure). Clearly, this increases the processing overhead required to execute not only the functional applications (the actual functionality desired by the systems integrator), but also the software infrastructure that implements and controls the VCs and TCs. However, with the significant increase in the available processing power of modern microprocessors, this should not be a significant problem assuming that an analysis of the total processing requirements is undertaken before the computer is procured.

For 'Plug and Play' weapons integration within the ASAAC architecture, a Run-time Blueprint would be developed for each stores load-out as part of the integration process. A 'Plug and Play' Store Configuration Domain may be responsible for discovering the load-out and requesting any required mode transition. In that mode, software processes are assigned to computing resources, software threads are dispatched, data access is permitted, communication between threads is scheduled and so on. Other non-functional aspects of the system may also be configured by a Blueprint such as safety and security partitions, and the re-allocation of resources in the event of hardware failure.

Blueprints are used to configure the allocation of resources such that the level of performance required to implement the functional Applications can be achieved and the Application Manager (in the AL) enables new application processes to be correctly scheduled at run-time. The Blueprint is a file which is encapsulated in the compiled computer programme; it exists as a part of the operational system software. When a new Application needs to be implemented (e.g. when a new weapon function capability is to be introduced), the Blueprint can be modified to share out the available resources (e.g. memory partitions) such that an optimum level of system performance can still be achieved (assuming that the underlying computing element still has sufficient processing power available). It is the GSM (residing in the OSL) which actually interprets the Blueprint in order to configure or re-configure the system according to actual needs, allocating the necessary resources as required. This level of flexibility in the ASAAC architecture makes a significant contribution to the integration of new functional capability.

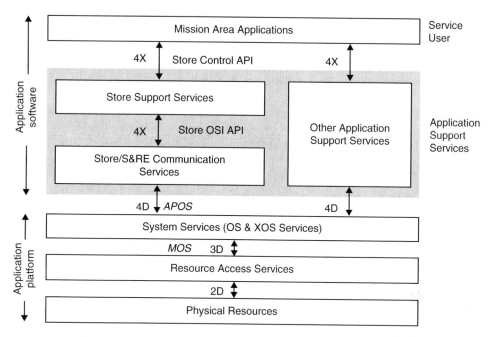

Figure 12.5   ASAAC instantiation of 'Plug and Play' implementation

Figure 11.2 identified the GOA profile for an aircraft 'Plug and Play'-compatible computer. Figure 12.5 reproduces this figure as it would relate to the overlay of the 'Plug and Play' implementation on an ASAAC-type architecture. It can be seen that the only tangible difference is that 'API' has been replaced by 'APOS' and the MOS interface is now shown.

In order to understand how a 'Plug and Play' Domain could be implemented within the ASAAC architecture, the following example will explore a possible implementation for part of the Store Control Service. In this example, a generic ALWI-CI API derived communication subset is mapped to the ASAAC APOS software calls.

By adopting the MDA® modelling approach, the Domain models would handle any required API call. For example, considering the weapon used in Chapters 9 and 10, the XPGB may need to receive an 'XPGB.Arm' command to initiate the weapon's thermal battery. In relation to the implementation of the MIL-STD-1760 interface, this would require 28VDC2 to be energised, Release Consent to be enabled, and the correct data messages transmitted over the MUX Bus. Within the software architecture, this API call may require the services of other domains. Therefore, the XPGB.Arm function call could be split into several other function calls (or operations) that are handled by other classes within the overall model, even if they are located in other Domains. These lower level function calls might include operations such as PwrSupply.Activate28VDC2, SignalProcessor.ActivateReleaseConsent and XPGB.SendArmCommand. The XPGB. SendArmCommand might also require a communication service, whose domain class usage could include a command to the MIL-STD-1760 domain to send data to the store, for example, 'Station 5: Receive Message 11, 30 words' over the 1760 MIL-STD-1553 bus hardware.

12.8   Certification Issues

With current system qualification and certification methodologies, whenever a system is modified, it is not uncommon for much of the system to be completely re-certified. Except for very small changes to the system, the bulk of certification costs (including the costs of re-qualification) are re-spent in generating the new clearances. For complex systems, there will always be a fixed portion of the re-certification cost and a variable portion, with the cost of certification being loosely related to the size and complexity of the system, as shown in Figure 12.6. As typical mission systems become larger and more complex, certification costs rapidly increase. Ideally, re-certification costs should be proportional to the size and complexity of the system change with the fixed portion of costs being kept as small as possible as shown in Figure 12.7.

With legacy systems, integrity is assured by reliance on a proven development process as the means of generating evidence for the safety case. During development of the system, the certification process is applied with the aim of ultimately demonstrating that the total system

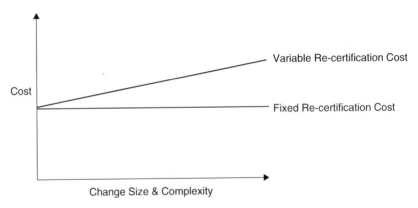

Figure 12.6   Cost of re-certification related to system size and complexity

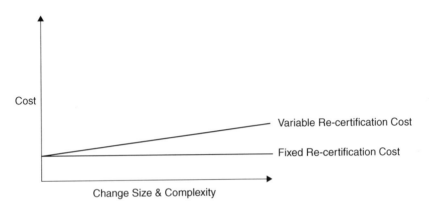

Figure 12.7   Cost of re-certification related to size and complexity of change

is safe. If the system software is subsequently upgraded (to integrate new weapon functionality), the safety case evidence for the original system is invalidated, requiring the whole system to be re-analysed to generate the new safety case evidence. This affects the whole system and not just the new and modified parts from the original system. In effect, the upgraded system is largely treated as a totally new system. Little evidence can be read across from the earlier system standard to support the new safety case. The upgraded system is demonstrated to be acceptably safe, but this is time-consuming and so incurs high costs. Therefore, the actual cost of certification is broadly related to the complexity and size of the system.

An alternative method of certifying the system is to identify the properties of the system which are deemed to make it safe. By analysing and qualifying these aspects, a safety case can be constructed that demonstrates the system to be acceptably safe. When the system is subsequently upgraded, the unchanged properties are imported into the new safety case argument.

In order to gain benefits with this approach, there is a strong case for considering the partitioning of the system and software architecture at the start of development, such that the desired level of modularity required in the safety case can be assured from the outset. For example, the SMS implementation may vest integrity in a relatively simple but tightly specified (e.g. using formal mathematical methods to define the requirements) interface controller. In this example, the interface controller would generate safety-related weapon interfacing signals when appropriately demanded by the mission system. A different mission system computer would contain the mission-critical weapon control functionality. With this architecture, a new weapon would be integrated by the addition of mission system AL applications only, leaving the underlying layers (OS and MSL) unchanged. It is, therefore, only those parts of the system software affected by the upgrade that need to be re-analysed. The complete safety case for the upgraded system therefore consists of evidence from the original system parts that have not been changed, plus the new evidence generated for those parts of the system which have been affected. In this example, only those parts of the system affected (i.e. the new applications and any other software applications which interface with the new/modified applications) need to be scrutinised in detail. Therefore, the fixed portion of the re-certification cost is significantly reduced with the additional re-certification costs becoming proportional to the size and complexity of the change rather than to the size and complexity of the total system. Clearly this reduces the certification effort.

Pragmatically, a hybrid approach of the two methodologies would be used. However, the hybrid approach may be biased towards the process viewpoint rather than considering only the parts of the system which have been changed. Therefore, in order to reduce the costs of re-certifying an upgraded system, more emphasis must be placed on the qualification of the product properties. The move closer to this alternative methodology means that incremental certification can be realised, thereby improving the cost relationship in line with the ideals of Figure 12.7.

When considering a system which has implemented a 'Plug and Play' capability, two sets of data would be generated. One set would be developed at the PIM phase with the other developed at the PSM phase to configure the run-time system. As these layered software architecture components would be proven at their design time, certification of a new weapon load-out would be achieved largely by validating the two new data sets.

12.9   Easing the Upgrade Programme

This chapter has shown that there are a number of steps that move from systems that are expensive and time-consuming to upgrade to an open systems approach. However, it will be rare for a new open system to be designed from scratch, particularly for legacy aircraft programmes. It is more likely that elements of open system and modular application concepts will be introduced as upgrade programmes allow (e.g. when re-designing a computer in order to overcome component obsolescence issues). The level of intrusion into existing systems will be dictated by customer requirements and the needs of the ADO to be able to manage risk as more flexibility to upgrade and simplify the integration of new weapons is demanded. In moving from legacy system implementations towards modular systems presents a challenge to find the correct balance of software architecture modularity and intrusion of re-work to existing application processes. However, once a truly modular architecture has been implemented, there are increasing paybacks in terms of application portability, processing resource allocation and standardisation.

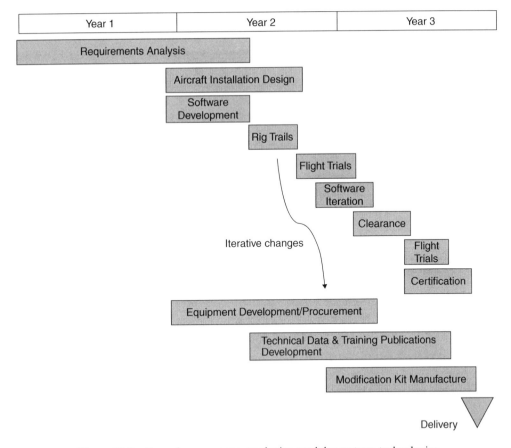

Figure 12.8   Upgrade programme employing modular systems technologies

Although this level of modularisation gives significant benefits to system upgrades, it should be noted that this is not the final solution for 'Plug and Play' weapons integration. However, it is an important step on the journey as it does provide the basis for an aircraft system that would be used to implement standard weapon applications such as Transfer Alignment and inventory management. The modular applications architecture does ease the addition of new weapon applications. This is directly relevant to the adoption of standard weapon control services as defined for a 'Plug and Play' capability. All the necessary weapon-specific applications would reside in the AL and would be configured by weapon-specific data.

In Figure 12.2, the typical mission system upgrade programme was explored. If modular systems architectures are to be introduced that use standardised hardware and implementing common industry standards such as ASAAC, then many of the original activities are reduced in magnitude leading to the initial system being certified sooner, and the time taken to produce further system capability improvements drastically reduced, depending on the level of intrusion necessary. Figure 12.8 shows how the programme could be reduced by the adoption of such technologies.

This chapter has explored enabling change in a mission system by the development and practical realisation of modular architecture open systems. This is an essential step in easing the weapons integration programme. The adoption of modular applications supported by the three-layer software architecture, de-coupled from the underlying hardware, enables the adoption of design for upgrade principles. These concepts also enable a move towards incremental certification. The result is an aircraft system that can be easily upgraded with new weapon applications or hardware and certified in reduced timescales. The level of standardisation required also leads to the potential for re-use/portability of applications. This provides a major building block in simplifying weapons integration where weapon-specific applications can be easily added. The benefit for the end user is that new weapons can be integrated into the mission system sooner and at lower cost. The ability to manage programme risk and system integration costs, coupled with the Contracting Agency's need for incremental upgrades to maintain force flexibility, satisfies the needs of the weapons integration business.

Further Reading

Society of Automotive Engineers. (1996) *Generic Open Architecture*. Aerospace Standard AS4893, Society of Automotive Engineers, Warrendale.
United States Department of Defense (2007). Aircraft/Store Electrical Interconnection System. *Military Standard 1760*, US Department of Defense, Philadelphia.

13

The Universal Armament Interface

13.1 Chapter Summary

The United States Air Force has identified that the time taken to field new weapon capabilities is intrinsically linked to the aircraft Operational Flight Programme (OFP) procurement cycles. Traditionally, should a new weapon requirement be needed, integration could only commence at the start of the next OFP upgrade cycle. As such cycles may only occur every three to four years, this has the potential to significantly delay the introduction to service of new weapon capabilities. The response to this situation has been the development of UAI, a concept which aims to break the dependence on the OFP upgrade cycles, thereby enabling new weapons to be introduced to service quickly. This chapter will outline the UAI concept and the potential benefits it brings to weapons integration programmes.

13.2 Introduction

As has been noted in previous chapters, a lack of standardisation between the aircraft and the store makes each integration programme unique. This means that every aircraft has to provide the store-specific interfaces and aircraft software has to be modified to provide the required functionality. In Chapter 11, it was noted that one of the recommendations from the early ALWI studies postulated that stores integration could be simplified by adopting a standard interface approach for all stores covering physical, electrical and system functional interfaces. Whilst clearly it would be impossible to define a standard interface that captures all the possible interface needs for current and future stores, the adoption of common physical and electrical standards (such as MIL-STD-1760) has provided a significant contribution to simplifying

Aircraft Systems Integration of Air-Launched Weapons, First Edition. Keith A. Rigby.
© 2013 John Wiley & Sons, Ltd. Published 2013 by John Wiley & Sons, Ltd.

integration. The early ALWI studies postulated a greater level of standardisation of the ICD. It is this 'common ICD' approach which has been developed in the United States as UAI.

The objective of a common ICD is to standardise the interface between stores and aircraft and to eliminate the practice of supporting multiple unique interfaces within the aircraft. Once an aircraft implements the common ICD standard, it would have the capability to integrate all weapons that implement the common ICD without having to change system software. Consequently, weapon integration costs could be substantially reduced. In addition, the amount of time required to integrate a weapon would be significantly decreased since software specification, development, coding and overall system validation would also be significantly reduced. Core certification activities such as assessing aircraft handling in its new configuration, aeromechanical aspects such as verifying that the aircraft environment does not have a detrimental effect on the store, safe-separation activities, guided firings and so on would still be undertaken as required.

This chapter will provide an overview of a common ICD approach as realised in UAI and then introduce the NATO initiative to standardise UAI internationally.

13.3 Objectives of UAI

Whenever a new store is to be integrated with an aircraft, it is usually assumed that this will be a stand-alone programme. However, this is rarely the case. In reality, a weapons integration programme is part of the through-life capability improvements of the platform. In addition to introducing a new weapon capability, other programmes could be introducing other capability such as a new sensor, general software functionality improvements or introducing new equipment that overcomes component obsolescence. Therefore, a real aircraft through-life capability programme consists of a disparate set of individual capability programmes which are the subject of separate contracts from the Contracting Agency.

Figure 13.1 shows a typical set of integrated capability programmes and a parallel weapon development programme.

Here, we see a typical software upgrade programme in progress during the first six years of the overall time period shown. This could be introducing system enhancements or other new capability to the aircraft. It is assumed that during the first year of this programme, the software requirements for this new capability need to be frozen such that the update can be developed and tested, allowing for a correction cycle before the new capability is released to service. In this example, the development of a new weapon commences at the end of the first year. This is completely independent of the aircraft upgrade programme. When developing a new weapon, it is usual to identify the lead platform for the integration programme (the aircraft being considered in this example). As the aircraft in this example is already in an upgrade programme when the weapon development programme commences, and the engineering team which will integrate the weapon are employed discharging this programme, there is no opportunity to feed in new requirements without delaying the fielding of the new capability scheduled for service release at the end of year 6. Also, as the weapon is in early development, its integration requirements are likely to be evolving, meaning that it would be too early to commence an integration programme with manageable risk. Therefore, platform integration on the lead platform cannot commence until a level of weapon design maturity has been achieved (towards the end of the weapon's development programme). In this example, the

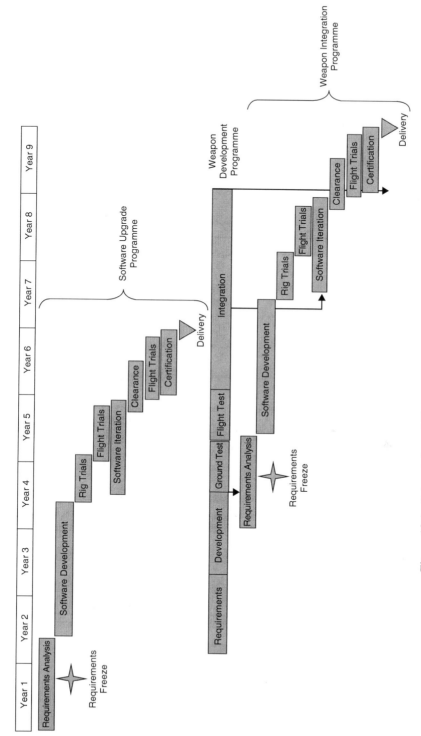

Figure 13.1 Integrated capability programmes with parallel weapon development

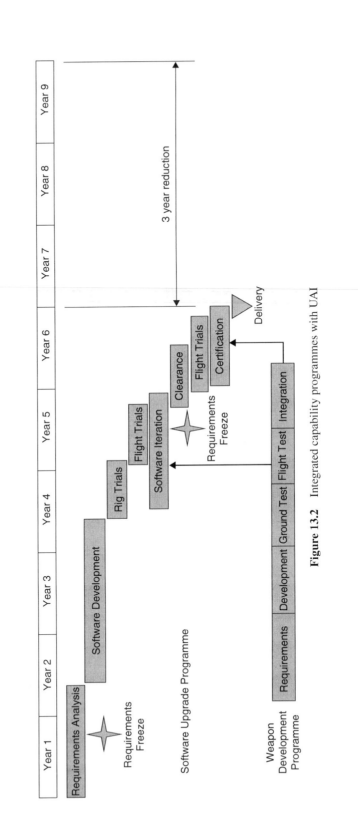

Figure 13.2 Integrated capability programmes with UAI

integration programme for the weapon is ready to commence at the start of year 6 with requirements from the weapon development programme having been provided early in the aircraft programme's requirements analysis phase. The aircraft integration programme then follows the typical route of software development, testing and iteration where a final set of requirements are captured from the weapons integration programme. This example demonstrates that for a real aircraft programme which undergoes through-life capability enhancements, the total elapsed time from commencing weapon development to fielding the new capability on the lead platform is of the order of eight years.

In the previous example, the integration programme is intrinsically linked to the software upgrade cycle of the aircraft. The main objective of UAI is to decouple the functional integration of new weapons from the aircraft update cycle. In doing so, the implementation of UAI involves a one-time activity to update an aircraft to provide standard interfaces suitable for all weapons within specific classes. This upgrade is performed independently of a specific weapon type. However, for UAI to be successful in realising a reduction in integration timescales, the weapon must also be designed to comply with the UAI standard interfaces. Once UAI-compliant aircraft and weapons are available, the subsequent functional integration programmes will be significantly reduced (potentially to weeks/months). However, such savings can only be made once the aircraft has been upgraded to be UAI-compliant, and for legacy aircraft, this is likely to incur a significant cost and timescale on its own. When referenced to the overall savings accrued by integrating multiple weapons throughout the aircraft's life, it is postulated that significant savings will be made. However, with the perceived high cost of the initial system implementation of UAI in the aircraft, the real cost benefits will only be accrued if a number of new weapons are integrated throughout the remaining service life of the aircraft.

Figure 13.2 shows an example of the potential benefits of adopting a common ICD solution such as UAI.

The 'common ICD' nature of UAI (which will be explained in the following sections of this chapter) enables a rapid integration of the weapon with the aircraft. When integrating a new UAI-compliant weapon with a UAI-compliant aircraft, as the weapon development programme nears completion (when the systems interface is finalised), the systems integration requirements can be fed in to the next available software iteration cycle. This level of flexibility can therefore translate into a significant reduction in the time required to integrate the weapon with the aircraft. In this example, this could be as great as three years, meaning that the new weapon's capability can be fielded sooner and for a significantly reduced integration cost.

13.4   Fundamental Principles of UAI

UAI defines a protocol for achieving the rapid integration on weapons with a combat aircraft. The UAI standard is based around the MIL-STD-1760 Class II interface but requires the aircraft's computing architecture to provide a common interface within the OFP software that is configurable without changing the software. As it addresses the system aspects of weapon integration, UAI does not define any requirements relating to the physical/aeromechanical integration.

The initial version of UAI is optimised for the integration of Air-to-Ground smart weapons although there are no fundamental reasons why this could not be extended to cover other smart

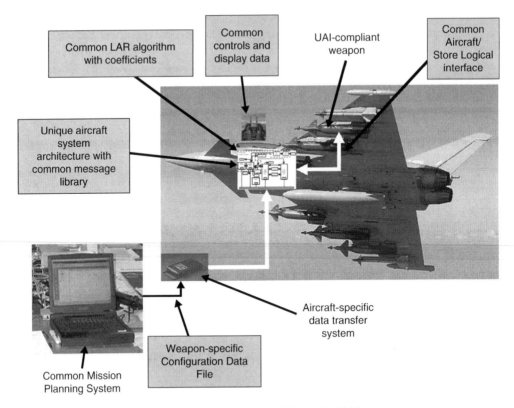

Figure 13.3   Interfaces addressed by UAI

weapons such as Air-to-Air missiles or weapons complying with interface standards such as AS5725 or AS5726, which themselves employ the MIL-STD-1760 data protocols and logic.

The UAI documentation defines a common Aircraft/Store Logical interface (based on MIL-STD-1760), common data for controls and displays, a common LAR algorithm configurable by data coefficients, a common message library and a common Mission Data interface.

Figure 13.3 shows the 'common ICD' approach adopted by UAI. This requires the aircraft to implement the full message and functionality set defined, with the store able to use the relevant subset of messages and functionality required to support its functions. However, different aircraft will have their own unique system architectures and legacy technology. UAI is designed to be independent of these aspects, enabling the functions and performance (e.g. the rate at which data parameters are generated) of the specific aircraft system to be accommodated.

The actual subset of messages and functionality employed for a particular UAI-compliant store has to be defined and the aircraft subsystems configured to provide this. Therefore, a key aspect of UAI is the use of Configuration Data Files which are used to configure a receptive aircraft system. This means that new weapons can be introduced to the aircraft inventory without the need to change system software, thereby decoupling integration from the OFP upgrade cycle. The Configuration Data Files are generated by UAI-specific software components embodied in the Mission Planning System and include sufficient information to configure the LAR algorithm and provide data relating to the aircraft displays. Of course, the data generated by the Mission Planning System also contains other data required by the aircraft such as route plans, target data and so on. This is transferred from the Mission Planning System using whatever means is provided by the aircraft (e.g. a data transfer cartridge) although the data formats are standardised using MiDEF (see Chapter 7). The exact content of the Configuration Data Files required for a particular aircraft/store combination are generated during the weapons integration programme. This is explained in more detail in Section 13.8.

13.5   Platform/Store Interface

The ALWI-2 study recommendation, which was the catalyst for UAI, recommended the adoption of a common ICD as a means of improving interoperability and to reduce the effort, and therefore timescales associated with integrating a compliant weapon with a compliant aircraft. The initial definition of UAI was applied to direct attack Air-to-Ground weapons adopting the MIL-STD-1760 interface.

UAI is based on the premise that all stores in the same class have similar interfaces. That is, all weapons in the same class generally require the same kind of data exchanges, have similar employment timelines and require common data of similar resolution, integrity and update rates.

By employing the MIL-STD-1760 interface, much of the logical data sets required by direct attack Air-to-Ground weapons are already provisioned. As explained in Chapter 6, the standard does include some standard messages but leaves most as being free for the weapon designer to use as required. Whilst this feature of MIL-STD-1760 provides a great deal of flexibility, it also means that dissimilar logical interfaces will be implemented by different stores. UAI, taking a 'common ICD' approach, strengthens the definition of the logical interface such that there is significantly less scope for 'user-defined' messages. The UAI Platform–Store ICD (PSICD) provides a comprehensive set of logical functions including discovery (which enables the aircraft and store to discover each other's capabilities), a standardised method of Transfer Alignment, GPS initialisation and so on. The standardised UAI message set must be implemented by the aircraft with the store using a subset of those messages.

In order to facilitate UAI, the MIL-STD-1760 logical set includes a number of features specifically required to support greater interoperability (e.g. Interface Configuration ID – used to allow upgrading of a standardised logical interface, while supporting backward compatibility).

Although the PSICD is called an ICD, it is really an Interface Specification with individually identified requirements which are traceable within the overall aircraft and store designs. The UAI documentation set also includes a Validation Plan which is used to test aircraft and stores for compliance. The Validation exercise ensures that every UAI requirement that has been implemented has been implemented correctly and in a manner that is compliant with the standard and therefore interoperable. An aircraft or a weapon is validated against the requirements of UAI by the use of bespoke Certification Tools and not by compliance to the interface of the component (weapon or aircraft) with which it is to be integrated. This means that once an aircraft has been proven to be compliant to the PSICD, then by definition, any store that has been proven to be compliant with the PSICD is also interoperable with the aircraft. This is a powerful feature of UAI. However, it should be noted that the aeromechanical aspects of weapons integration (e.g. aeromechanical modelling, safe-separation analysis, etc.) will still need to be performed.

As the aircraft is required to implement the full UAI PSICD but the store only required to implement the subset of the ICD required for its operation, there needs to be a method of configuring the interface such that the aircraft systems know exactly the logical set, timelines and so on required by the store. This is done as a part of the integration programme as described in Section 13.8.

13.6   Mission Planning

The flexibility provided by the implementation of UAI in aircraft and stores means that the aircraft systems need to understand relatively little about specific stores prior to store discovery. Clearly, during the discovery process, the aircraft system needs to determine what functionality it needs to provide to control the store. That is, it needs to understand which parts of the PSICD need to be initialised. To maintain this flexibility and avoid the need for hard-coding of store functionality in the aircraft system, the required UAI configuration for a particular sortie is configured using data generated by the Mission Planning System. This data will include the usual mission data (route plans, target information, etc.), but for UAI, the Configuration Data Files for the aircraft/store combination must also be generated. As UAI was developed in the United States, it relies on the use of a common Mission Planning System – the Joint Mission Planning System.

As mission planning for all aircraft employs common functions (e.g. route planning, attack planning, etc.), a common Mission Planning System will use a set of core software modules to generate data which can then be formatted for the specific aircraft data transfer system. The implementation of UAI into a Mission Planning System can be achieved by introducing specific software components. Figure 13.4 shows how UAI would be implemented in a common Mission Planning System.

In order to support UAI, the common core modules are supplemented by common UAI mission planning modules. These modules provide all the specific files required to config-ure UAI. For a Mission Planning System that has to be compatible with a particular aircraft, an aircraft-specific Unique Planning Component (UPC) is loaded. This enables the mission planner to access the correct aircraft UAI configuration data. When a new weapon is to be integrated with the aircraft, a Weapon UPC is required which defines its configuration information. Using the Aircraft UPC and Weapon UPC, the specific Configuration Data Set can be generated and packaged together with the remainder of the mission plan. The result-ing mission plan data is then loaded on to the aircraft-specific data transfer medium so that it can be loaded into the aircraft system. In order to maintain interoperability, all the data required to be transferred to the aircraft employ MiDEF. In order to control the UPCs, the UAI documentation set includes a Mission Planning ICD defining all the required parameters.

Although UAI mission planning is based on the US Joint Mission Planning System, there should be no fundamental reason why UAI mission planning modules cannot be developed and integrated in any Mission Planning System based on a commercial operating system.

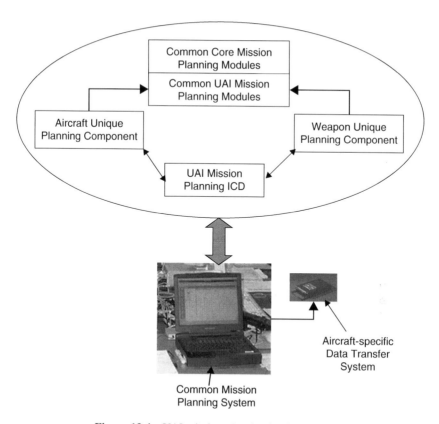

Figure 13.4 UAI mission planning implementation

13.7 Launch Acceptability Region

Chapter 7 introduced the CLARA documents (AIR5682 (CLARA ICD), AIR5712 (CLARA Rationale Document) and AIR5788 (CLAR Truth Generator ICD for the CLARA). UAI has adopted this approach and defines the processes required to support the management of LARs.

The CLAR approach is based on having a standard algorithm which can be configured by data coefficients. The WDO is responsible for developing the LAR Weapon Truth Data which defines the actual aerodynamic performance of the store. As a part of the integration programme, the WDO provides the Truth Data to the ADO for inclusion in the Configuration Data Set. The ADO also implements the basic LAR algorithm in the aircraft system, but in order to ensure that the performance of the weapon can be achieved, there is also the need to analyse the accuracy of the LAR. This is documented by the ADO and included in the documentation set used for certification of the integrated system including the new weapon.

13.8 Integration Work Flow

When a UAI-compliant store is to be integrated with a UAI-compliant aircraft, as noted earlier, this can be achieved outside the normal aircraft software upgrade programme. Figure 13.5 shows the integration flow.

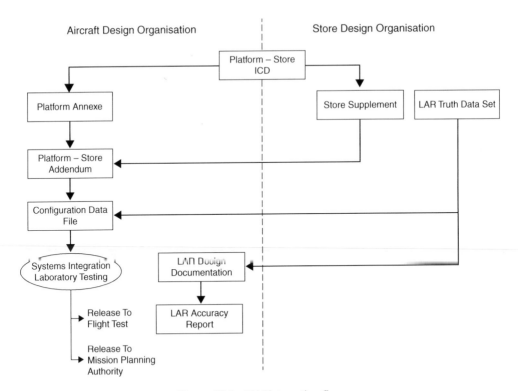

Figure 13.5   UAI integration flow

The actual logical interface adopted by the store is documented by the WDO in a Store Supplement. The Store Supplement identifies all of the PSICD requirements which have been implemented to provide the required interface functionality. It is therefore the responsibility of the ADO to configure the aircraft interface to match the needs of the store.

The ADO produces a complementary document to the Store Supplement known as the Aircraft Annexe. Together, these two documents form the Platform–Store Addendum, and from this, the Configuration Data File for the aircraft/store combination is developed. For an aircraft that can operate several UAI-compliant stores, there will be separate Configuration Data Files for each combination. All the files are combined into the Configuration Data Set which is used to configure the aircraft system. In effect, it is the separate Data Files which sets the operational capability of the aircraft.

The aircraft/store combination will then be tested in the SIL to check for correct operation of the stores being integrated. Any system integration problems which are identified can be corrected simply by an iteration of the configuration data. Once the system is deemed to be working as required, the final documentation set can also be baselined.

When SIL testing has been satisfactorily completed, the final version of the Configuration Data File for the aircraft/store combination is released for flight test. If during the flight test programme further system integration problems are identified, then these can be corrected purely by modifying the configuration data.

Although the overall test regime can still be significant, dictated largely by the complexity of the store and the need to develop a level of confidence appropriate for certification, it can be seen that the cost and timescales associated with integration are greatly reduced. This is due to the systems integration task largely consisting of generating data files.

The final Configuration Data File is also released to the authority responsible for the Mission Planning System such that the new functionality can be implemented.

UAI has been demonstrated as a rapid means of integrating a new store with an aircraft, with the time taken to have a system configured to enter SIL testing reduced to months/weeks or even days. Although there will still be a need for the aeromechanical activities to be undertaken and the overall aircraft/store combination certified, the reduction in systems upgrade timescales provides a significant cost saving on the overall programme.

As noted previously, the initial cost of UAI implementation into an aircraft can be very large, particularly for a legacy aircraft, meaning that through-life cost benefits can only be achieved if a number of store integrations are to be undertaken. In the United States, where the armed forces have a large number of aircraft and weapon types, then the business case for adopting UAI is clear; although it should be noted that not all weaponised aircraft will implement UAI, as the business case would be difficult to justify for aircraft with a relatively short in-service period remaining. Building UAI into the baseline for new aircraft would be easier to justify and would provide the best through-life savings. However, for nations which do not have the number of aircraft and weapon types found in the United States, the business case for adopting UAI may be more difficult to justify.

13.9   UAI Interface Management

Chapter 8 outlined the process adopted in the United States for managing interfaces between the aircraft and the store. This takes a store-centric approach, based on integration programmes for an individual store with multiple aircraft, where the ICD covers primarily the store, but with aircraft annexes defining the specific interface implementations. As UAI is, by its very definition, supposed to be compatible with a multitude of different aircraft and stores based around the common ICD documentation set, then the store-centric approach is not tenable.

Figure 13.6 shows the UAI Super JICWG, where the focus is on management of the UAI ICD's, annexes, supplements and addenda. It is noticeable that the construct of the Super JICWG is very similar to the construct of the JICWG shown in Figure 8.2. However, as there are several ICDs to manage, then the management organisation is duplicated with Chairpersons, Interface Managers and Participants being appointed for each of the three UAI ICD documents (the PSICD, the LAR ICD, and the Mission Planning ICD). These are supported by a body of advisors, and there is a common Secretariat to ensure consistency of process and administration across all ICDs.

As described in Chapter 8, the Super JICWG is also controlled overall by the Interface Management Board with a higher level of management (the Interface Control Board) available to arbitrate in the resolution of any interface control issues raised by the Super JICWG which cannot be satisfactorily settled by the Interface Management Board.

The Interface Managers and Participants of the Super JICWG are drawn from the various programmes (Aircraft 1, Aircraft 2, etc., and Store A, Store B, etc.) which are party to UAI. Clearly, this leads to a large number of people from different organisations being involved in the control and development of UAI, which itself leads to a complex management problem. However, the principles adopted by UAI are consistent with the US interface management process defined

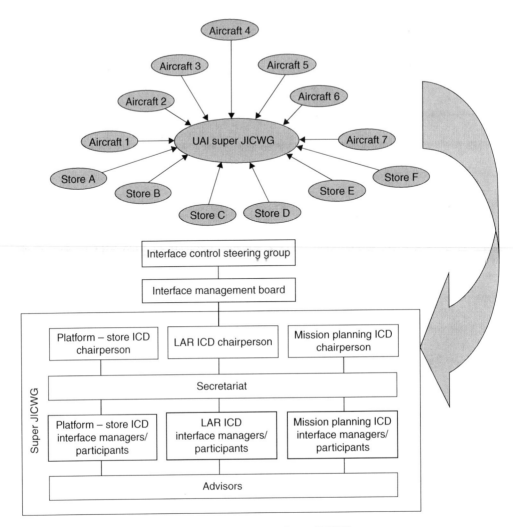

Figure 13.6 The UAI Super JICWG

in Chapter 8. To facilitate the management of UAI, the roles and responsibilities of all parties have been defined in an Interface Control Plan.

13.10 Certification Tools

As noted previously, UAI departs from the established principle of proving that the aircraft and the store correctly implement the specific interface requirements of the other. With UAI, each component (aircraft and store) is independently validated against the UAI PSICD using a standard Certification Tool based on commercially available test equipment, integrated to test all features of the logical interface. The PSICD Certification Tool provides a versatile test capability which can be configured to validate the correct implementation of UAI for Mission

Stores and Carriage Systems independently of any aircraft with which they may be integrated. The same Certification Tool can also be used in a SIL to validate an aircraft's implementation of UAI. It is this configuration which is the most useful to the ADO, as the tool contains a number of test scripts which can fully exercise the aircraft system. For example, the Certification Tool can be configured in its simplest form as a UAI-compliant Mission Store. However, it can also be set to operate as a Mission Store with variable timing constraints which are close to the acceptable limits defined by the PSICD. This feature enables the aircraft system performance to be explored for all allowable variations.

The tool can also be configured to act as a Carriage Store loaded with up to eight weapons, taking into account the Carriage Store architectures allowed by UAI (these are a simple Type 1 Carriage Store, where the aircraft system maintains full direct control over the weapons and directly interfaces with the Carriage System S&RE; and a more complex Type 2 Carriage Store, where the weapons are isolated from direct aircraft control by a smart system in the carrier).

As UAI is dependent of the Mission Planning System, there is a need to also certify the UAI aspects of the Mission Planning System software. The Mission Planning Certification Tool therefore has to provide a suite of test software that can exercise the UAI Mission Planning components (the UAI Common Component, the Platform Unique Component, and the Store Unique Component). Each test component is required to exercise the component under test by simulating each of the other components. For example, when testing the UAI Common Component, the test software has to simulate the Platform and Store Unique Components.

By providing a set of standardised Certification Tools, any implementation of UAI can be validated for compliance. Once validation is complete and, for example, the aircraft has been demonstrated to be UAI-compliant, any UAI-compliant store (Mission Store or Carriage Store) can be integrated purely by defining a new Configuration Data File. This brings with it an unprecedented level of interoperability far in advance of previous standardisation efforts.

13.11   Benefits

The original aim of UAI was to enable 'out of cycle' integration of weapons with an aircraft. Practical experience of implementing UAI has shown that a significant benefit can be realised in terms of reduced integration timescales, leading directly to a reduction in systems integration costs.

UAI changes the way in which engineers approach a weapons integration programme. As a UAI-compliant aircraft contains all the functional capabilities required to control compliant weapons, the emphasis is shifted to selecting the required functionality rather than developing an ICD; defining the requirements for a configuration file rather than defining new OFP software requirements; and developing the configuration file rather than coding new software. By having aircraft and stores that have been certified as being UAI-compliant before the integration programme commences, system test and validation are also significantly reduced.

Published conservative estimates of the savings to be made once an aircraft has been modified to be UAI-compliant suggest that the time required to field a new weapon can be reduced by 25% compared to fielding a new non-UAI weapon. For capability upgrades to existing weapons, a 60% reduction in time is forecast. All this is postulated to translate into a 30% saving in integration costs.

A further benefit of UAI is that once aircraft, stores and Mission Planning Systems comply with the UAI ICDs, there is the opportunity to integrate new capability more frequently. This is likely to be attractive to military users as the utility of aircraft can be improved as new mission requirements are identified, for example, if an aircraft is needed to support urban warfare operations with weapons with reduced yield warheads.

13.12   NATO UAI

A prime objective of NATO is to have a greater level of interoperability across nations' equipment. In the aircraft/store domain, this translates to an ideal that any weapon from any nation can be operated by any aircraft from any nation, albeit with a level of reduced performance (this was the original view of increased interoperability identified by the original ALWI study). UAI is totally consistent with this aim, and it was the ALWI-CI study which recommended that UAI should provide the basis for a NATO Universal Armament Interface (NUAI). However, it was recognised that UAI may need to be extended to cover both aircraft and, more specifically, weapon variations across non-US NATO nations.

In order to progress towards NUAI, NIAG Sub-group 142 was initiated to review UAI and to make recommendations of essential changes required to the UAI documentation set that would be needed to encompass non-US stores. NATO also requested the SAE to develop a Super JICWG process which could be adopted for NUAI. This has now been published as AS6030 (Interface Standard, Common Interface Control Plan).

Whilst NATO can put in place all the building blocks required for NUAI, it is national Governments which procure systems. Even with the increased level of interoperability that would be available with NUAI, there is no guarantee that it will be adopted across all NATO nations unless sound programme business case justifications can be made.

13.13   'Plug and Play' Conclusions

This chapter has given an overview of UAI and outlined its benefits. UAI is clearly a 'Plug and Play' methodology which allows compliant stores to be integrated with compliant aircraft when desired by the military rather than being dependent on complex and costly software upgrade cycles. However, whilst UAI has been implemented on a number of US programmes, it should be noted that the cost of upgrading an aircraft to be UAI-compliant is likely to be significant as all subsystems which contribute to controlling a weapon must be modified together. Therefore, although weapons integration programme costs can subsequently be reduced, it will be necessary to integrate several weapons to recoup the initial upgrade costs. This means that UAI may not be the answer to reducing integration costs and timescales for all aircraft programmes, and it will therefore be essential to understand the effect on through-life weapons integration costs to determine if UAI is the correct solution for a particular aircraft programme.

Chapter 11 outlined the model-based, data-driven 'Plug and Play' approach developed as a concept under the ALWI studies and separate from the development of UAI. Whilst this version of 'Plug and Play' has the potential to be introduced piecemeal as aircraft upgrade programmes dictate, it will still be costly to implement the full capability. However, whilst accepting that the full benefits of 'Plug and Play' will not be realised until all the aircraft's subsystems involved in controlling weapons have been modified, the through-life business case could be more attractive to some nations.

Due to the investment in UAI made by the United States, this capability is now relatively mature, albeit with a need for continued development of the concept to encompass the integration of a wider set of store classes. UAI is also the basis for NUAI, which is aimed to foster greater interoperability across NATO nations. By comparison, the ALWI 'Plug and Play' concept is still relatively immature, although several European nations are continuing research and development in this area.

When considering which 'Plug and Play' technologies should be adopted, there will be a need to determine if some or all of the ALWI 'Plug and Play' concept should be adopted or whether it would be more beneficial to adopt NUAI. The UAI 'common ICD' concept does not detail how systems should be modified to implement the enabling flexibility, meaning that proprietary system designs within the aircraft can be retained. The ADO would have to devise the best method of implementing UAI for their specific programme. The ALWI 'Plug and Play' concept will already have provided the solution to this. However, ALWI 'Plug and Play' has the potential additional benefit of providing a greater level of flexibility than the minimum required for the implementation of NUAI; for example, it would provide a far greater opportunity for the integration of non-NUAI stores.

Further Reading

Society of Automotive Engineers. (2005) CLAR Truth Generator ICD for the CLARA. *Aerospace Information Report 5788*, Society of Automotive Engineers, Warrendale.

Society of Automotive Engineers. (2007) Common Launch Acceptability Region Approach Interface Control Document. *Aerospace Information Report 5682*, Society of Automotive Engineers, Warrendale.

Society of Automotive Engineers. (2008) CLARA Rationale Document. *Aerospace Information Report 5712*, Society of Automotive Engineers, Warrendale.

Society of Automotive Engineers. (2011) Interface Standard, Common Interface Control Plan. *Aerospace Standard 6030*, Society of Automotive Engineers, Warrendale.

United States Department of Defence. (2007) Aircraft/Store Electrical Interconnection System. *Military Standard 1760*, US Department of Defense, Philadelphia.

United States Department of Defence. (2007) Mission Data Exchange Format. *Military Standard 3014*, US Department of Defense, Philadelphia.

14

Weaponised Unmanned Air Systems

14.1   Chapter Summary

The advent of weaponised unmanned air systems (UAS) brings new challenges for weapons integration. Several programmes have fielded so-called 'killer drones'. This chapter will consider the differences between weaponised manned and unmanned systems and outline strategies for partitioning the overall system between airborne and ground-based segments.

14.2   Introduction

Present-day conflicts are increasingly employing unmanned airborne systems for the collection of imagery and intelligence. With increased persistence over the area of interest, such systems (often referred to in the media as 'drones') provide military commanders with an unprecedented surveillance capability. This capability has also been extended to include weaponisation, enabling commanders to prosecute targets quickly, thereby reducing the 'sensor to shooter' timeline. It is these combined capabilities, particularly with the employment of precision weapons, which have made the weaponised Unmanned Air Vehicle (UAV) a weapons system of choice. UAVs do not replace manned aircraft in all scenarios as there are times when 'eyes on target' in real time is essential to satisfy rules of engagement and to enable action within the laws of armed conflict.

The removal of the pilot from the aircraft means that UAVs of different weight classes can be designed. UAVs with purely a surveillance role can be hand-launched by infantry soldiers. Weaponised systems can be as large as existing manned aircraft, but could also be significantly smaller, carrying a few light-weight weapons such as the example depicted in Figure 14.1.

Legal requirements dictate that the employment of weapons must always be sanctioned by a human. However, the Ground Control Station for the air vehicle could be located thousands

Aircraft Systems Integration of Air-Launched Weapons, First Edition. Keith A. Rigby.
© 2013 John Wiley & Sons, Ltd. Published 2013 by John Wiley & Sons, Ltd.

Figure 14.1   Example of a small weaponised UAV. Reproduced by permission of BAE Systems

of miles from the theatre of operations. Therefore there is a need to transmit commands and receive status information over a data link which will have uncertain availability, delay and integrity. This significantly increases the complexity for the safe control of weapons, which needs careful consideration during the system design phase.

Depending on the complexity of the UAS (which will, as a minimum, consist of an air vehicle and a ground segment), the distribution of the system components between the ground segment and the airborne segment becomes a choice for the system designer. Work undertaken by NIAG has proposed various options for partitioning the system.

This chapter will explore the different design considerations for the weapon control system and review possible system partitioning options.

14.3   Distributed Weapon System

The obvious difference between a manned aircraft and a UAS is the removal of the pilot from the air vehicle, placing the operator at the ground segment (this could be a Ground Station located at the launch and recovery base, at the air force headquarters, on-board a naval vessel, in an airborne command and control platform, etc.). The weapon control functionality will therefore need to be distributed between the airborne and ground segments. The data links used to connect the two segments of the UAS can either be by a line of sight link, a beyond line of sight link (e.g. via a satellite) or a combination of both. Line of sight data links tend to have relatively low transmission delays but could be subject to the link being temporarily disconnected due to physical issues such as terrain masking, antenna masking during platform manoeuvres and so on. Beyond line of sight data links can suffer

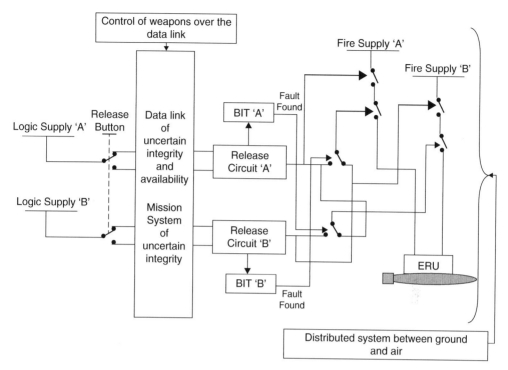

Figure 14.2 Distributed SMS Architecture

from a relatively limited data bandwidth and excessive transmission delays. The data link delays and the fact that the pilot is removed from the air vehicle lead to a loss of situational awareness. This can have significant implications for compliance with the rules of engagement and the laws of armed conflict.

Figure 4.1 showed an implementation of the SMS on a manned aircraft. This implementation satisfied the criteria that no single fault shall cause a release of a store when not intended, and no single fault shall prevent a store being released when intended. Figure 14.2 shows how such an implementation would be affected for a UAS. Here, the SMS is distributed between the air segment and the ground segment of the overall system.

With no human occupant in the vehicle, the rules of engagement have to be interpreted by the operator at the ground segment using situational awareness data that could have been transmitted from the air vehicle several seconds earlier. Weapon release will be initiated from the ground segment using the Release Button, with the release demand being transmitted over the data link to the air vehicle. In Figure 14.2, both the ground and airborne elements of the SMS design maintain the 'no single failure' design principles. However, this cannot be maintained across the data link which will be a generic, bi-directional link used to transmit air vehicle command and control data and sensor imagery between the two segments. The link, if it uses communications satellites, will introduce a significant delay to the receipt of the weapon release commands from the ground segment.

If integrity of the end-to-end control of weapon release is to be maintained, then it is likely that complex data protocols will be required. The end-to-end system will also need to implement strategies for maintaining synchronisation of the distributed system during data link interruptions.

Having a distributed system will complicate the certification activities, as although it is possible to control the design and implementation of the two segments, the two segments can be connected together by a multitude of possible data links employing different technologies, wave-bands, data rates, delays and so on, over which the ADO may have little or no control.

14.4 System Architecture Partitioning

The previous section considered only the distributed SMS and not all parts of the system which contribute to the control and deployment of weapons. Depending on the complexity of the weaponised UAS, the total functionality required to control and release weapons can be distributed in different ways across the ground and air segments. The NIAG Sub-group 125 study into UAS weaponisation postulated three levels of partitioning of the system functions.

The partitioning chosen will define the data which has to be transferred over the ground-to-air data link. Placing more functionality in a complex air vehicle can simplify the Interface Exchange Requirements (IERs) for the data link; a simpler airborne system may require more direct control of the weapon control functionality from the ground, employing a different set of IERs.

Figure 14.3 shows a weapon control Domain Model relating to the control and deployment of weapons of a weaponised UAS, and Table 14.1 provides an explanation of each of the Domains.

Generally, the complexity of a UAV is dependent on its mission capability which is linked largely to its weight class. Clearly, the more capable the sensor package, the longer the air

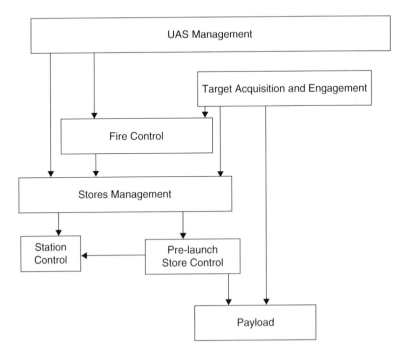

Figure 14.3   Weapon control system Domain Model

Table 14.1 Domain model components

Domain name	Function
UAS Management	This domain is responsible for the management of the UAS components during the mission (both in flight and on the ground). It includes controls required to operate the air vehicle, a means of displaying sensor imagery and operator controls for the armament system (selecting weapons for release, initiate jettison, initiate release, etc.).
Target Acquisition and Engagement	This domain is responsible for target acquisition and allocation of a target set to a weapon package. This will be implemented in the air vehicle's Mission System, controlling sensors (possibly in a multi-sensor turret containing visible/IR wavelength sensors) and providing functions such as laser designation for weapons which require it.
Fire Control	This domain computes the LAR and/or release cues for the selected weapon package.
Stores Management	This domain manages the carriage, release and jettison of all stores on the air vehicle. It manages weapon inventory and status. The domain provides services that contribute to the airworthiness of the vehicle/weapon combination, system safety and weapon safety.
Station Control	This domain manages the individual store stations on the air vehicle. Each store station provides S&RE for the payload, with the interfacing to the S&RE (e.g. rack firing) provided by this Domain.
Pre-launch Store Control	This domain manages a single mission store, possibly via a carriage store. For example, providing the MIL-STD-1760 interface functionality for the weapon.
Payload	The payload may comprise a single mission store or a specific mission store on a carriage store.

vehicle can be airborne, and the larger the mass of the weapon load-out, the larger the vehicle will be. This means that the aircraft mission system is likely to be more capable, enabling more of the weapon control functions to reside within the air vehicle systems. For a small UAV such as the example shown in Figure 14.1, the on-board mission system capability is likely to be reduced, meaning that more of the overall UAS functionality relating to weapon control will reside in the ground segment.

The NIAG UAS Weaponisation study defined three potential levels of air vehicle capability, identified as Level 1, Level 2 and Level 3. These are shown in Figures 14.4, 14. 5 and 14.6.

Clearly, there will be some Domains of the overall UAS which will always be a part of the ground segment such as the UAS Management Domain, but the other Domains may be partitioned to either the air vehicle or the ground segment.

Figure 14.4 depicts a system partitioning with the minimum of system elements located in the air vehicle. With such a vehicle, the weapons (payload) are carried on S&RE, and assuming that they employ an intelligent interface such as that provided by MIL-STD-1760, AS5725, or AS5726, there will be a need to control the weapon directly, providing target data, Transfer Alignment and pre-launch control commands and so on. The functions allocated to the air vehicle are therefore Station Control and Pre-launch Store Control.

For a Level 1 Capability air vehicle, the Stores Management and Fire Control Domains are partitioned to the ground segment. The inventory is managed in the Ground Station, and

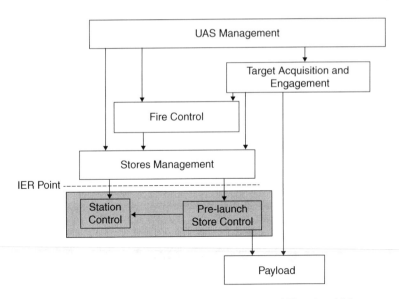

Figure 14.4   UAS partitioning for a Level 1 capability air vehicle

when a target is to be engaged, the actual weapon to be released is selected by functionality residing in the Ground Station. Similarly, the Fire Control Domain is partitioned to the ground segment, where functions such as LAR calculation or the computing of weapon release points will be performed and then transmitted to the air vehicle to be actioned. A potential problem with computing the LAR/release cues in the Ground Station is that the data link transmission delays from the Ground Station to the air vehicle will need to be accurately predicted to ensure an accurate release of the weapon can be effected. This can be extremely difficult, particularly when the end-to-end data link could consist of several hops using both line of sight and satellite communication systems, creating a potentially widely variable transmission delay. Therefore, a method will be required for predicting when to issue the release cue, such that when it is received and actioned by the airborne system, the weapon is released at the correct time.

The point in the system where the IERs for the data link are defined is also shown in the figure. NATO has introduced a level of standardisation for UAS data link interfaces by the introduction STANAG 4586. Whilst this standard provides a data word/data bit level of definition for information transferred over the Ground-to-Air data link, the work of NIAG Subgroup 125 defined abstracted data functions. IERs for the Level 1 Capability partitioning include interface requirements for both the Station Control Domain (e.g. S&RE Status, Store Release/Jettison, etc.) and the Pre-launch Store Control Domain (e.g. Store Status, Weapon Initialisation, etc.) with the Stores Management Domain.

A Level 2 Capability air vehicle (shown in Figure 14.5) places more of the UAS functionality in the airborne segment.

Besides the Station Control and Pre-launch Store Control Domains, the Stores Management Domain is migrated into the air vehicle. In doing so, the airborne system is required to provide a greater level of control of the weapons. This enables the operator in the ground segment to request a particular weapon type to be selected. The Stores Management Domain will select

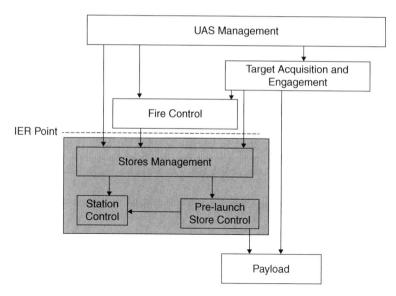

Figure 14.5 UAS partitioning for a Level 2 capability air vehicle

the actual weapon to be released, sequence the necessary timeline implementation through the Pre-launch Store Control and Station Control Domains and manage the update of the overall stores inventory post release. Should the weapon fail to be released, then the Stores Management Domain will remove the weapon from the inventory of stores available for release and notify the ground segment that a hung store is present. The Stores Management Domain also provides the required level of weapon system integrity.

The IER point is now in a position above the Stores Management Domain. IERs for this level of partitioning include requirements for the UAS Management Domain (e.g. Build Stores Inventory, Make Stores Management Live, Select Jettison Package, Release Weapon Package, etc.), Fire Control Domain (e.g. Select Weapon Package, Store Status Data, Release Weapon Package, etc.) and Target Acquisition and Engagement Domain (e.g. Weapon Package Data for Target). In a system partitioned with a Level 2 Capability, the UAS Management, Fire Control, and Target Acquisition and Engagement Domains provide the same functionality as they would in a Level 1 Capability system.

A Level 3 Capability air vehicle (shown in Figure 14.6) places the majority of UAS weaponisation functionality in the airborne segment. The air vehicle now contains the Fire Control Domain responsible for selecting the weapon package and the relevant release parameters and calculating release cues/the LAR. It is the Fire Control Domain which controls the release point of the weapons package and is the IER point for the ground-to-air data link. The Capability 3 system partitioning employs similar Stores Management IERs as the Capability 2 system partitioning. However, these will be modified, as the interface to the Fire Control Domain in a Capability 2 system partitioning becomes an internal interface within the airborne system for a Capability 3 system partitioning.

The additional IERs include data requirements for the UAS Management Domain (e.g. Weapons Initialised, Select Weapon Package, Release Weapon Package, etc.) and the Target Acquisition and Engagement Domain (e.g. Laser Designator Locked on Target).

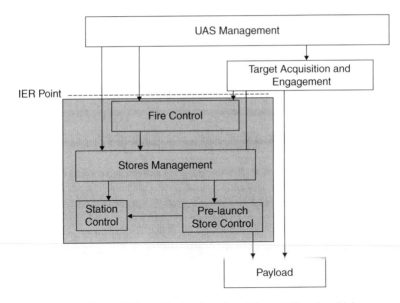

Figure 14.6   UAS partitioning for a Level 3 capability air vehicle

14.5   Conclusions

The distributed implementation of the weapon release functionality shown in Figure 14.2 reflects how the high-integrity elements of weapon control could be implemented in a UAS. It can be seen that this could be how a Capability 1 system partition would implement the Station Control Domain for a single weapon station. For a Capability 2 or Capability 3 UAS partitioning, the release circuits (Station Control Domain) would still be required for each weapon station, but would be supplemented by more complex SMS functions required to manage the inventory, select specific weapons into a weapon package, control release sequences and so on.

By adopting standard system partitions such as described in Section 14.4 enables the ground-to-air data link IERs to be standardised, thereby enabling standard protocols to be defined for different classes of UAS within STANAG 4586. However, defining IERs by understanding partitioning at the Domain level means that no single system implementation is mandated, leaving it to the ADO to determine exactly how the system is to be implemented and, therefore, which parts of the STANAG 4586 interface messages are to be adopted.

Further Reading

NATO Standardisation Agency. (2012) Standard Interface of UAV Control System for NATO Interoperability.
    STANAG 4586, NATO Standardization Agency, Brussels.

15

Reducing the Cost of Weapons Integration

15.1 Chapter Summary

Both industry and national procurement organisations agree that the costs associated with the integration of a new weapon with an aircraft are high and the time taken to field new capability is too long. The cost drivers are many, and for costs and timescales to be reduced, there is a need to address a multitude of factors. This chapter will identify the cost drivers and the initiatives required to improve business efficiencies such that weapons integration programmes can be streamlined, thereby reducing integration costs. The chapter will also argue that even in a climate of reducing defence budgets, efficient weapons integration programmes can be a source of increased business for the ADO.

15.2 Introduction

The previous chapters of this book have explored the changing face of aircraft systems integration of air-launched weapons. Air-launched weapons are becoming more capable in terms of the range from which they can be deployed and the accuracy with which they can hit their target, and provide the effects demanded in modern warfare. However, the nature of warfare is constantly changing, with lower priority now being placed on the kind of total war that was envisaged during the Cold War. Today a range of differing scenarios exist which encompass large-scale conflicts and lower intensity conflicts such as asymmetric warfare, defence against terrorism and operations confined to relatively small urban areas (the so-called 'three-block war') as well as peace-enforcement operations. Military aircraft can have long service lives. As an example, the Interdiction/Strike variant of the Tornado aircraft (the GR1 variant) entered service with the air forces of the United Kingdom, Germany and Italy in the early

Aircraft Systems Integration of Air-Launched Weapons, First Edition. Keith A. Rigby.
© 2013 John Wiley & Sons, Ltd. Published 2013 by John Wiley & Sons, Ltd.

Figure 15.1 Tornado GR4 with typical weapons load-out. Reproduced by Permission of Geoffrey Lee, Planefocus Limited

1980s and is planned to be in service until around 2020 (for the Royal Air Force's GR4 variant). Changing military needs mean that the weapons capability of a particular aircraft will continuously change throughout its lifetime, with new weapons being integrated. With the Tornado example, the original weapon inventory included free-fall and retarded bombs, cluster munitions, practice bombs, and short-range Air-to-Air missiles. During its service life, the weapon inventory has been supplemented with a range of other smart stores including Paveway™ guided bombs, anti-radiation missiles, the Storm Shadow stand-off missile and Brimstone short-range Air-to-Ground missiles (Figure 15.1).

The trend is for weapons to have smaller warheads which are capable of being delivered with extreme precision. These new weapons will have to be integrated with the aircraft systems in order that the overall combined system performance can be optimised.

The through-life costs of maintaining cutting-edge weapons capability are significant. National Governments are increasingly realising that to control the cost of weapons integration and to provide value for money for tax-payers, whilst delivering the capabilities demanded by the war-fighter, a closer collaborative engagement is needed between the Contracting Agency, the ADO and the WDO. There is also a need to transform behaviours, the structure of the various organisations and the business models of all parties in order to have better controlled efficient programmes.

In Chapters 11, 12 and 13, efforts aimed at making significant reductions in the cost and time-scales for introducing new weapon capabilities to aircraft were explored. Although two approaches (ALWI 'Plug and Play' and UAI) were covered, both were the result of addressing the need to ease the systems integration aspects of air-launched weapons integration. The key reason why

NATO originally promoted the need for 'Plug and Play' weapons integration was to introduce a greater level of interoperability across member air force's aircraft, launchers and weapons. As a pure military need, there cannot be any guarantee that industry will drive this initiative forward unless a clear business imperative can be identified. In order to understand why defence contractors became involved with the development of such technologies as ALWI 'Plug and Play' and UAI, it is first necessary to understand the weapons integration cost landscape.

15.3   The Cost Landscape

Actual programme costs are extremely hard to determine as, often, the integration of a new weapon with an aircraft is part of a larger programme. Also, detailed figures are deemed to be commercially sensitive and are therefore not usually published. The actual cost can also vary depending on a number of factors. For example, if a weapon is in development, there will be a need to undertake flight trials as a part of proving that its implementation meets the specification. If the nation developing the weapon wishes to employ an aircraft type with which the weapon will eventually be integrated (usually the lead platform) to support the weapon development programme, then it is not unusual for the Contracting Agency to let two contracts on the ADO. These will be a contract to support the development of the weapon, with a separate contract to cover the integration of the weapon with the aircraft to provide the required in-service capability. By employing two separate contracts, the apportionment of actual costs between weapon development and the aircraft/weapon integration programme become blurred.

If the weapon is effectively off-the-shelf and requires no further development, then the programme to integrate it with an aircraft is simplified as only the aircraft systems will need to be modified. However, the relative complexity of the weapon and the aircraft will have a strong bearing on the actual cost of integration. When integrating a highly complex weapon on a 'simpler' aircraft, the changes to the aircraft systems may be significant, needing new equipment to be introduced and possibly significant modifications introduced to existing equipment and software. Clearly, the difference in complexity of the aircraft and the weapon could have a significant impact on actual integration costs.

NATO, national Contracting Agencies and industry have investigated the actual cost landscape, and whilst few figures relating to actual weapon integration programmes have been published, the combined wisdom across a ten-year period from around 1996 to 2006 suggests that a typical aircraft/weapons integration programme would equate to a cost of between 40 and 150 million US dollars.

These figures are for the complete integration programme, but again, there is a level of confusion as to what is covered and what is not. For example, if through-life support costs are included, then the cost will be significantly increased. For a programme where training costs are excluded, then the cost could be under-estimated. Therefore, in order to understand the apportionment of costs purely to the systems integration aspects of the programme, a number of assumptions need to be made. For the sake of this discussion, it is assumed that the elements of the programme which make up the cost landscape are systems engineering integration and test (including all the activities covered in Chapters 2 and 3), development of aircraft software, airframe integration, development of mission planning software and miscellaneous aspects such as training and technical publications.

Figure 15.2 shows the typical breakdown of the integration costs for a MIL-STD-1760 precision-guided munition. It is assumed that the aircraft requires no hardware modifications, and existing mission planning equipment is used.

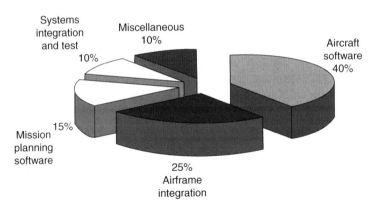

Figure 15.2   Typical MIL-STD-1760 aircraft/weapon integration Cost breakdown

From the figure, it is clear that the largest component of the programme's cost is apportioned to the aircraft's software development life cycle. The development life cycle for the mission planning software is also a significant cost. When added together with the systems integration and test activities, these comprise almost two-third of the total programme costs.

The initiatives discussed in Chapters 11, 12 and 13 are all aimed at reducing the costs apportioned to aircraft system software (around 40% of the total costs). The means by which these costs may be reduced for a system/software based on the ALWI 'Plug and Play' concept (notwithstanding the variables which can, themselves, have a significant effect on the percentage of integration costs attributed to software changes) is discussed later. In the discussion, it is assumed that the establishment of the MDA® tool environment, any certification activities associated with that tool environment, the initial development of the Domain architecture, process improvements and training of engineering personnel in the methodology have already been the subject of previous investment.

When considering the systems integration of a new weapon, then only changes to the software would be required. As the ALWI 'Plug and Play' concept has not yet been fully developed and fielded, there are no published cost metrics on which to determine the benefits. However, it is not unreasonable to suggest that any cost savings would be based on the avoidance of software design, coding and test effort. In the ALWI 'Plug and Play' concept, the development of source code for the upgraded capability would be replaced by activities to develop data files, plug-ins and so on. Industrial experience with the use of MDA® in other industries suggests that a 25%–30% improvement in software productivity would be achieved over the conventional software development life cycle. In the case where new weapon capability is introduced using an existing store class, the changes required to the aircraft system software would be reduced further. In principle, it is estimated that the software costs could be reduced by the order of 40%–50%.

When looking to justify any investment in 'Plug and Play' technology, whether this is ALWI 'Plug and Play' or NUAI, then there will be a need to understand the benefits of the technology during the total life of the aircraft. National Governments may have military capability road maps which can be translated into the need for new weapons. However, these are, by their very definition, a forward-looking view based on the anticipated future threats. Whilst such road maps are rarely released to the public, information gleaned from press releases, formal

Government statements and so on can give some insight into a nation's future intentions. For example, a strategy in the United States to develop smaller weapons became apparent when SAE International was requested in an open meeting to develop AS5725 and, later, AS5726. In the United Kingdom, the need for the Selected Precision Effects At Range (SPEAR) capability was publicised in the House of Commons Hansard Written Answers for 30 Apr 2002 (column 743 W).

Although a forward-looking view can shed light on the potential future weapons integration business, there is also merit in looking at previous weapons integration programmes that have been completed, compared to what was originally planned over the same period. This was a method used in the United States to justify the business case for the development of UAI, where an analysis showed that significantly more stores had been integrated over a ten-year period than had actually been planned at the start of the same period. Therefore, the business case for the adoption of any new technology aimed at reducing the cost and timescales of weapons integration needs careful consideration of both the future planned programmes and a retrospective look at actual past programmes to determine the potential future business for the ADO and WDO.

In the United States, there are generally more programmes largely due to the number of aircraft types and a greater range of weapons. However, the trends are for fewer types of more capable weapons. In contrast, European nations have fewer aircraft types, and due to consolidation in the industry, a smaller range of weapon programmes. This results in significantly fewer weapon integration programmes in Europe than in the United States.

A relatively small number of infrequent programmes mean that key weapons integration skills may not be maintained within the national industrial base. The response to this situation is for national Governments to promote the need for a greater level of export business to offset the need for spending at home. If successful, this strategy can increase the throughput of programmes, thereby maintaining key skills and increasing profitability for the industrial organisations involved in the programme. However, the dichotomy is that to secure export programmes, integration costs must be affordable for the export customer, particularly as such programmes are usually the subject of significant international competition. If weapon integration programmes are deemed to be unaffordable, this can have a direct impact on the ability for nations to export both aircraft and weapons. Therefore, the added imperative is for integration costs to be reduced, but also for the timescales to be carefully managed such that new capability can be fielded with the export customer in as short a time as practical.

15.4   Reducing the Cost of Weapons Integration – Other Initiatives

The general acceptance that the cost and timescales of typical weapon integration programmes is too high is a significant issue for the Contracting Agency, the ADO and the WDO. Whilst the adoption of systems technologies can have a significant bearing on reducing the overall cost of a stores integration programme, by concentrating on this area only will not reap all the potential benefits. Whilst this book has not addressed the aeromechanical aspects of integration (which, from Figure 15.2, accounts for around 25% of integration costs), clearly this is an area which could also benefit from new technology and integration processes. There are also a number of other areas which, if addressed, can provide a significant contribution to making the overall programme more efficient. The key is to ensure that only activities which are either necessary to ensure certification or add value to the programme are undertaken.

From the perspectives of the ADO and WDO, there is also a need to ensure that a viable business enterprise is maintained. Clearly, from the WDO's viewpoint, this is achieved through greater sales of their products. For the ADO, this is achieved by ensuring that new capability is introduced to their products throughout the in-service life of the platforms. However, where a stores integration programme differs from other capability programmes is that there is a need to work closely with another DO as well as the Contracting Agency. This tri-partite interaction introduces a complex set of interrelationships which can often be based on trying to satisfy the different needs of the stakeholders. It is this which brings the opportunity to identify other initiatives for increasing the efficiency of the programme and thereby reducing integration costs.

In order to ensure that the ADO in particular continues to have a viable weapons integration business, future integration programmes will dictate the need to adopt new business models. To achieve this, there are a number of key enablers as follows.

15.4.1   Streamlined Integration Processes

Streamlined integration processes will help to remove activities from the programme which do not directly add value to the integration of the store with the aircraft. To achieve this, there will be a need for the ADO, the WDO and the Contracting Agency to have a close working relationship such that all relevant stakeholder needs can be addressed at the outset of the programme. Relevant commercial constructs between the three organisations will be required to encourage open relationships. Such contracting arrangements may include provisions for sharing risks and rewards whilst allowing increased profit for industry in return for achieving reduced integration costs. This type of gain-sharing can be a powerful catalyst for streamlining integration processes.

15.4.2   Common Goals for the ADO and WDO

With closer relationships between ADO, WDO and the Contracting Agency, there will be the opportunity to understand different stakeholder priorities and develop common goals for the programme. Obviously, these will include the delivery of new capability to the war-fighter, but would also need to consider intra-programme goals such as defining common test requirements and generating shared certification data which can be used by the ADO, the WDO and the Contracting Agency's specialists for developing the new certification documentation for the aircraft/store combination. Developing intra-project goals could also lead to a greater level of inter-programme goals, where synergies could be realised across programmes when the store is to be integrated with several aircraft. For example, identifying the synergies across multiple programmes at the outset of the integration programme for the lead platform could mean that expensive test assets such as weapon simulators or environmental data gathering weapon variants could be shared across the various platform programmes. Where the store is also in its development phase, there will be the opportunity to define integration requirements that satisfy the needs of all the intended platforms with which it is to be integrated. By managing inter-dependencies between aircraft programmes more effectively and ensuring that the integration of a new weapon on the lead platform takes account of what follow-on platforms may need, integration costs for subsequent platforms may be controlled. Considering such inter-programme initiatives can have a significant bearing on the Contracting Agency's total costs of integrating a single store across multiple platforms.

15.4.3 Employment of New Technology Which Eases Integration

With defence equipment budgets coming under increasing pressure, national Contracting Agencies are required to deliver greater levels of capability at reduced costs. Streamlined integration processes and close working relationships will help to reduce the total cost of integration, but the time taken to upgrade legacy software code will become the dominant factor driving the overall programme timescales. As discussed in previous chapters, new technology such as ALWI 'Plug and Play' or UAI can have a significant effect on reducing the aircraft software element of the overall programme. Whilst this is likely to provide the single most positive benefit in reducing timescales and therefore costs, there can be other significant benefits. The adoption of such new technology will provide the means for more frequent, cost-effective updates to aircraft/weapon capability. Whilst this may not be directly relevant to a national Contracting Agency which is trying to operate with reduced budgets, the technology will provide a means for the ADO to integrate weapons quickly for export customers. This can provide the economies of scale which will provide a long-term business for the ADO, but also could be used to offset through-life capability costs for the national Contracting Agency, particularly if national and export requirements for new weapons capability can be aligned. Such industrial innovation and invention coupled with the adoption of international standards provides the mean to discharge competitive weapons integration programmes.

15.4.4 The Need for Exports

National Governments are increasingly relying on the economies of scale which can be achieved through export as a cost-effective means of securing new capabilities for their own forces. The ADO also relies on a throughput of stores integration contracts to maintain key skills. Whilst the Contracting Agency is primarily concerned with the delivery of new capability for home forces, the means by which the systems integration is accomplished will be determined primarily by the ADO. When considering how to modify the aircraft subsystems for a new store capability, the ADO should also consider the implementation trades which could ease the integration of future stores. By avoiding wherever possible, bespoke integration solutions, features which ensure that future store integration programmes can be achieved in a more cost-effective manner can be adopted. However, there will be a need for careful consideration of benefits when trading flexibility which supports future programmes, and the cost/timescales for providing the contracted capability. The Contracting Agency is rarely willing to pay the price for future flexibility for other customers unless there is a national strategy to support exports. If such a strategy exists, then there is the opportunity to investigate the potential synergies that could be achieved by understanding what the export market is likely to need when defining national requirements. Such a strategy can have a significant benefit for the overall through-life weapons integration costs for a particular aircraft programme, thereby reducing the cost to export customers, which, in turn, can increase competitiveness for the ADO. A through-life initiative adopted at the start of an integration programme can therefore provide a means of generating long-term value for money for the home Contracting Agency.

15.4.5   Spiral Introduction of Capability

When contracting for the integration of a new weapon, the end user may not have devised exactly how it is to be used in service. This can lead to User Requirements being defined which define the need for a greater level of flexibility and additional initial capabilities which, ultimately, may not be employed in service. This can directly drive the costs and timescales of the programme, which could be deemed to be too expensive or take too long to enter service. In the worst-case scenario, this could actually lead to the Contracting Agency cancelling the requirement for a particular aircraft type. To overcome this, there is a need to only define the essential requirements for the initial service capability. By taking this approach, the overall programme risk and timescales can be controlled thereby controlling the integration costs. Once the end user has had the opportunity to explore the new capability and to develop tactics for employment, further capability enhancements can then be considered and contracted separately. By adopting such spiral capability enhancements at the outset, the essential capability may be fielded sooner and for a reduced cost, whilst enabling cost-effective enhancements to be introduced to extend the initial capability. This approach will therefore deliver an initial capability that the war-fighter actually needs rather than what the war-fighter perceives they may need.

15.4.6   Organisational Re-structuring

A consequence of streamlining integration processes and adopting new technology which reduces the effort required to integrate a new store, coupled with a reduction in home demand and a greater dependence on the export market, means that both the ADO and the WDO cannot necessarily maintain the size of their organisations. This can drive the need for re-structuring to reduce the overhead costs carried by the organisations, thereby increasing efficiencies, which in turn can reduce integration costs of future programmes.

15.4.7   Adoption of International Standards

Standards provide a means of adopting common elements which can be re-used across multiple programmes. Generally, once a standard has been adopted, the effort required to integrate further stores which comply with the standard can be reduced. Standards continue to be refined to foster greater interoperability, and new standards are developed which are compatible with evolving military needs. These evolving needs, for example, the development on smaller munitions, will further stimulate weapons integration activity in the medium and long term. This will provide greater opportunities for the ADO to sustain business, but only if integration programmes can be undertaken within the cost and timescale aspirations of Contracting Agencies

15.5   Conclusions

Figure 15.3 (reproduced from Chapter 2) identifies all the systems integration activities (the white boxes) required to successfully integrate a weapon with an aircraft. Chapters 9 and 10 discussed the considerations for each of the areas relating to systems integration for a typical integration programme. A successful ADO will therefore need the competence to develop, deploy and maintain the knowledge and skills required to undertake these activities.

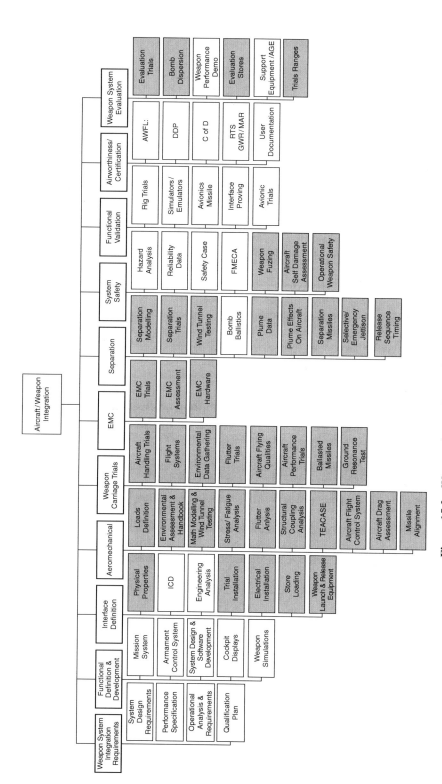

Figure 15.3 Weapons integration work breakdown structure

The many different skills required are diverse, and it is unlikely that the majority of experienced systems integration engineers within the ADO will have broad coverage of all these skills. Instead, an individual engineer may be expert in a few with a broader knowledge in several other areas. Very few, if any, will have detailed experience across all skill areas. For an organisation to develop and maintain a viable weapon system integration business, the collective capability of the engineering team must be expert in all the relevant skill areas.

A proficient ADO will have the ability to investigate how integration processes can be streamlined and new technology applied to generate efficiencies which improve the delivery of new military capability to the end user. The ultimate aim of the weapons integrators (within the ADO, the WDO, and the Contracting Agency) is to deliver an integrated aircraft and store that provides exemplar military capability for a competitive price and in the timescale needed by the war-fighter.

There are many opportunities for an ADO trying to secure profitable business. Being the design authority for a particular aircraft means that whenever a change is required to introduce modified or new capability, then a level of engineering activity is required. Simpler upgrades may be implemented by the end user although this means that from a certification viewpoint, the end user takes on responsibility for the safety and airworthiness of the aircraft. Whilst the ADO may work with the end user in the certification of service upgrades, it is the upgrades which are contracted to the ADO which really provide the attractive business opportunities.

System upgrades can take many forms and have an impact on different parts of the aircraft. Examples of these could be the introduction of modified equipment to overcome component obsolescence or to improve the reliability of system hardware. However, in order to exercise the maximum level of skills within the ADO, there are few capability upgrades which can compete with the potential level of complexity and diversity required when integrating a new store. Stores integration (and in the case of this book, the systems integration of air-launched weapons) can provide a significant through-life profitable business for an ADO which is able to manage efficient and cost-effective integration programmes. With the ability to generate new business with both the home Contracting Agency and export customers, there is the potential for a significant level of business. For example, analysing the likely future stores integration programmes for just one platform type throughout its service life is likely to reveal in excess of one billion dollars of potential business opportunities over a period of twenty years.

15.6   The Future

This book has concentrated on the systems integration of air-launched weapons and discussed trends which generally will be for smaller, smarter munitions. With the increasingly complex battlespace, these future weapons are likely to include greater network-enabled capability, employing data links, thereby allowing battle commanders greater flexibility in the prosecution and destruction of targets using physical weapons.

However, as technology advances, particularly in the area of high-power microwave and laser systems, new types of weapons, generally referred to as Directed Energy Weapons, are being developed. The UK Government's Defence Industrial Strategy of 2005, for example, stated: 'we judge that Directed Energy technologies could be highly significant in the future, particularly for protecting our Forces from a range of threats, including Improvised Explosive Devices. They could also offer the United Kingdom non-kinetic and/or less-lethal options to replace, enhance or complement traditional kinetic weapons, such as missiles, and offer significant opportunities to reduce collateral damage, notably in urban areas'.

Such weapons could be realised either as integral subsystems within an aircraft or as deployable stores/small UAVs. Where such weapons are deployable, then the skills required for integration will be largely as required for the integration of kinetic weapons. Clearly, where such a weapon is an integral aircraft subsystem, then from a systems integration viewpoint, many of the skills and capabilities resident within the ADO will still be relevant, but the focus will shift to provide a greater emphasis on weapon aiming, power scheduling and design for electromagnetic compatibility with other aircraft subsystems. The safety needs for the employment of Directed Energy Weapons will still drive the need for a safety critical control system.

The integration of these new types of weapon will no doubt bring new challenges for the weapons integrator. However, as many of the skills required for the systems integration of kinetic weapons will still be directly relevant to the integration of Directed Energy Weapons, the ADO must first become a proficient systems integrator before facing the challenge of adapting the processes and skills required for the integration of these novel weapon types.

Further Reading

Her Majesty's Stationery Office. (2005) Defence industrial strategy defence white paper, section B7.34, December, Her Majesty's Stationery Office, London.
United Kingdom House of Commons. (2002) Hansard Written Answers, April 30, column 743W, Hansard, London.

Index

Printed in the United States
By Bookmasters